CW00672308

Birds of Paradise

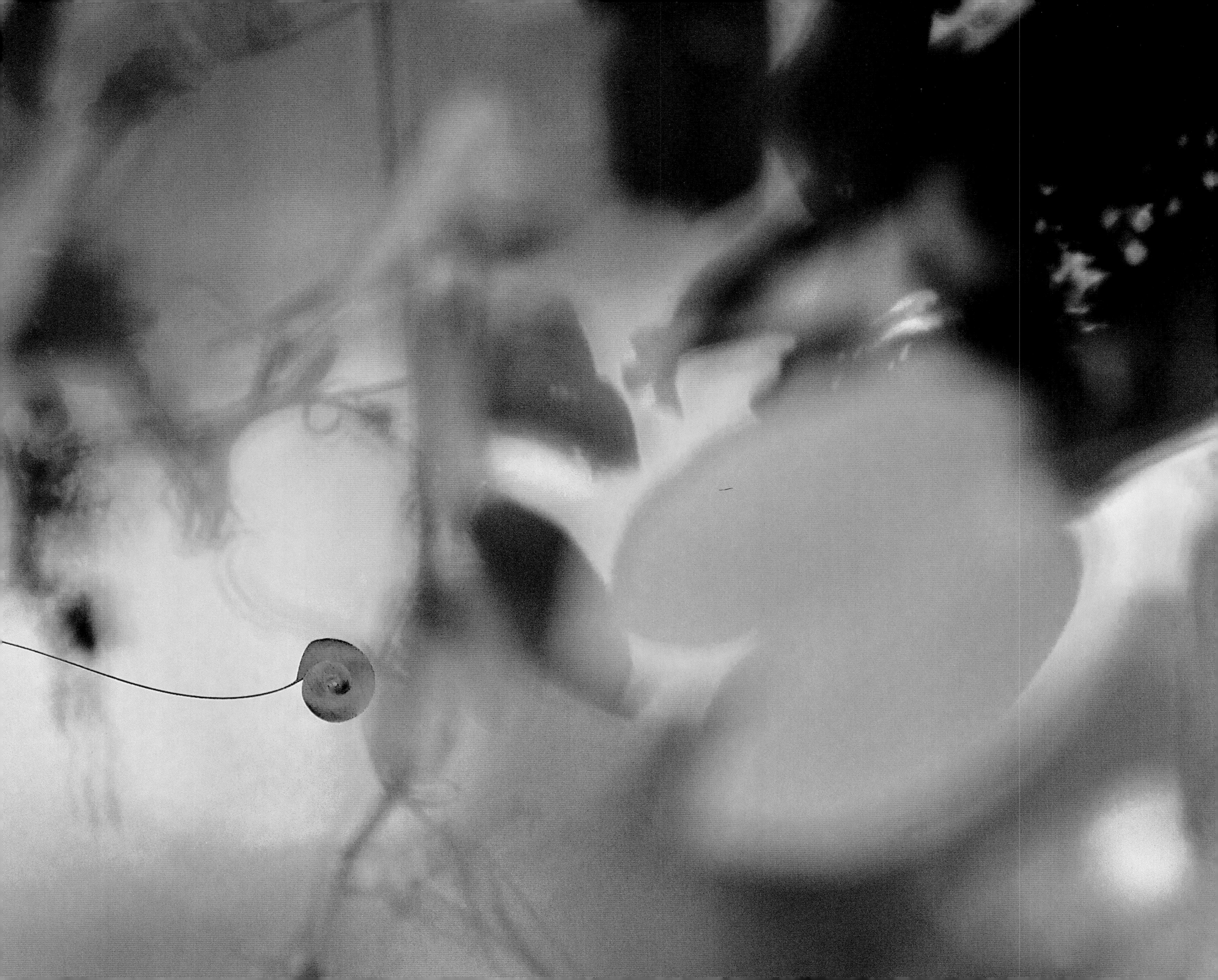

Birds of Paradise

REVEALING THE WORLD'S MOST EXTRAORDINARY BIRDS

Tim Laman & Edwin Scholes

NATIONAL GEOGRAPHIC | The Cornell Lab of Ornithology

WASHINGTON, D.C.

Contents

RED BIRD-OF-PARADISE ~ *Paradisaea rubra:* Wailebet, Batanta, 23 November 2004 *(right)*

KING OF SAXONY BIRD-OF-PARADISE ~ *Pteridophora alberti:* Mid Tari Gap, 8 December 2010 *(page 1)*

GREATER BIRD-OF-PARADISE ~ *Paradisaea apoda:* Wokam, Aru Islands, 21 September 2010 *(pages 2-3)*

KING BIRD-OF-PARADISE ~ *Cicinnurus regius:* Oransbari, Bird's Head Peninsula, 31 August 2009 *(pages 4-5)*

TWELVE-WIRED BIRD-OF-PARADISE ~ *Seleucidis melanoleucus:* Nimbokrang, Jayapura area, 14 June 2009 *(pages 6-7)*

SPLENDID ASTRAPIA ~ *Astrapia splendidissima:* Lake Habbema, Snow Mountains, 20 June 2010 *(pages 8-9)*

MAGNIFICENT BIRD-OF-PARADISE ~ *Cicinnurus magnificus:* Lower Satop, Huon Peninsula, 18 December 2006 *(page 10)*

Preface ~ The Birds-of-Paradise Project

Eight years. Eighteen expeditions. Fifty-one unique field sites. Thirty-nine species of birds-of-paradise, all photographed in the wild for the first time in history. This is what we have come to call the Birds-of-Paradise Project, an effort that has occupied a substantial chunk of our lives. This project didn't begin as a full-fledged plan expertly mapped to completion, but rather, like many difficult endeavors, it emerged slowly through a circuitous route and a series of fortuitous events.

It dates back to September 2003, when Tim Laman first emailed Ed Scholes to learn more about his National Geographic–funded field research on the birds-of-paradise of the genus *Parotia*. Tim introduced himself and informed Ed that he had recently been given an assignment to photograph the birds-of-paradise for an upcoming story for *National Geographic* magazine. By the end of January 2004, we met in person to discuss birds-of-paradise and the logistics of doing fieldwork, especially photographically oriented fieldwork, in New Guinea. Several days later, Tim made a proposition that ended up changing the course of both our lives forever and that led to the years of collaborative effort that ultimately led to this book. In an email dated January 26, 2004, Tim offered Ed the opportunity to accompany him as scientific adviser/photographic assistant for all the field expeditions for the *National Geographic* article. Ed gave an enthusiastic "Yes!" and the rest, as they say, is history.

By the end of 2006, we had completed seven major expeditions, Tim had photographed 22 species in the wild, and the magazine article was slotted to come out in the July 2007 issue. Ed had completed his Ph.D., and somewhere along the way, possibly in one of the many soggy field camps or on one of the many sweat-soaked hikes through the forest, a plan was hatched. Perhaps we had entered into a moment of heat-induced delirium, but for some reason, even after the challenges, we said, "Hey, we're halfway there!" and decided to try to raise the necessary funding to document *all* the currently recognized species. By that time, the species list had grown to a total of 39 with the announcement of the rediscovery of the Bronze Parotia in the Foja Mountains in early 2006. By late that year, we received continuing support from the Expeditions Council of the National Geographic Society, and we were officially off on our quest to document all the birds-of-paradise in the wild. We thought we would finish the work in another couple of years, but in the end, we required five more years and generous support from the National Geographic Society and the Cornell Lab of Ornithology, where Ed began working in early 2008, plus additional help from Conservation International.

Over the years, the project's goals also evolved. What began as an effort to photograph all the species with a book in mind grew into a larger effort to foster the advancement of knowledge through our scientifically based media collection (photos, video, and audio recordings) and to promote conservation and global appreciation of the New Guinea region through exploration, discovery, and education focused on the birds-of-paradise. In other words, our ambitions for sharing our results also grew as we started to put more emphasis on video in addition to still photographs.

As the project's scope grew, we also began to push the frontiers of how we could see and document the species. Tim continued his tree-climbing canopy work to see the birds in novel ways, but now he was also using multiple cameras in blinds and shooting both video and stills at the same time. The desire to push the boundaries didn't stop there, and as the years went on, so did the challenges. In 2010, Tim added a remotely controlled camera hidden in a bundle of leaves, a "leaf-cam," to his bag of photographic tricks. Now we were capturing images, both still and moving, of birds-of-paradise that showed previously unseen aspects of their world.

By the close of fieldwork in 2011, we had amassed the world's most comprehensive collection of scientifically gathered photographs and video and audio recordings of birds-of-paradise. All of the recordings were archived for research and educational purposes in the Macaulay Library at the Cornell Lab of Ornithology, where Ed is the curator of the Video Collection. The project had also grown in scope beyond just a book. It now included a large interactive educational museum exhibit, a documentary film, and a second feature article in *National Geographic*.

What started out as a shared fascination for the birds-of-paradise and our mutual interest in exploring the New Guinea region evolved into an unprecedented opportunity to fill a gap in the documentation of our planet's biodiversity. We turned our passion for science, photography, and media documentation into a comprehensive vision to share a treasure of the Earth's biodiversity with the world. We were lucky to be able to stand on the shoulders of naturalist-explorers such as Alfred Russel Wallace, Ernst Mayr, Tom Gilliard, Clifford Frith, and Bruce Beehler who had both come before us and worked alongside us. A project like this could never be completed without the contributions and support of so many. Our sincere hope is that this project will raise awareness about birds-of-paradise and New Guinea's rain forest and foster their stewardship.

We hope you enjoy the book. We know you'll love the birds.

—TL and ES

Tim works concealed inside a cramped canopy blind 40 meters (130 feet) high in a tree adjacent to a Lesser Bird-of-Paradise display tree.
Oransbari, Bird's Head Peninsula, 21 August 2009

Tim Laman

ON MY FIRST BIRD-OF-PARADISE photography trip, I wrote this field journal entry, dated 6 September 2004: "I think this is the most difficult assignment I have worked on so far . . . Tenth day since I arrived, and I still haven't nailed a single bird-of-paradise species to my satisfaction." I was not a newcomer to the rain forest or to photography, and on most assignments I would definitely be getting publishable photos by day ten. What had I gotten myself into?

In 1987, I had been inspired to work in tropical rain forests by reading Alfred Russel Wallace's *The Malay Archipelago: The Land of the Orang-utan, and the Bird of Paradise* (1869). The "Land of the Orangutan" referred to Borneo, where I gained my first tropical forest experience as a research assistant and then conducted my Ph.D. research over a number of years. A scientist at heart, but with an adventurous bent and a love of photography, I chose a doctoral project that allowed me to climb the big trees of Borneo's rain forest and explore the lives of strangler fig trees and associated wildlife high in the canopy. Because the project had great photographic potential, it was my training ground for rain forest photography as well as an exciting way to do science. It led to my Ph.D. from Harvard and also gave me my break at *National Geographic* magazine, where it became my first feature story in 1997. My varied interests were coming together into a career path—as a scientist/explorer documenting the natural world with a camera and sharing my stories with the world.

Throughout these early years, I dreamed of someday investigating the other part of Wallace's book title, the "Bird of Paradise." When my proposal to *National Geographic* to shoot a story on birds-of-paradise was accepted, my eight-year obsession began. By that time, I had a dozen published *National Geographic* features under my belt, including such other challenging subjects as the orangutans, hornbills, and gliders of Borneo. I knew covering birds-of-paradise would be difficult, but I felt ready for the task. Teaming up with ornithologist Ed Scholes was a key step. On that first bird-of-paradise trip in 2004, Ed had already displayed his exceptional field skills by locating the display site of a Brown Sicklebill, one of the least known birds-of-paradise. But after ten days into our first trip, I was realizing that this project was going to be even tougher than I had imagined. Eventually, though, we started to get some breaks. As soon as I started shooting some successful images of these amazing birds, I was hooked.

Edwin Scholes

THE EVENT THAT LED ME to the birds-of-paradise occurred 15 years ago on what was otherwise an unremarkable evening in Tucson, Arizona. Surprisingly, it began while watching television. At the time, I was nearing graduation with a biology degree from the University of Arizona and contemplating graduate school. I happened upon the opening credits of a film called *Attenborough in Paradise*, and the next hour changed my life forever. That documentary, the result of a lifelong ambition by Sir David Attenborough, was the first to capture the splendor of the birds-of-paradise in any depth. As I stared at the screen that evening, I found myself in a state of awe rarely attained from an armchair experience. What I saw filled me with wonder, curiosity, and desire. As the final credits rolled, I asked myself a simple question that has guided me since: "How did those birds evolve to be so beautiful, so diverse, and so bizarre?" As I sat there in the dark, illuminated by the blue glow of the television, I made myself a two-part promise. One, I had to see a bird-of-paradise in New Guinea with my own eyes, and two, I had to make use of my budding career in biology and a desire for science-based exploration to help answer the question I posed that night.

This book represents the partial fulfillment of that youthful promise. My 22-year-old self would never have imagined he would go on to complete a Ph.D. from the University of Kansas under evolutionary ornithologist Richard Prum (now at Yale) involving six years of fieldwork using video to study bird-of-paradise behavior in the forests of New Guinea. Nor would he have believed that he would go on to a postdoctoral fellowship at the American Museum of Natural History in New York City and land a job combining his interests in birds, evolution, behavior, videography, and natural history collections at the Cornell Lab of Ornithology in Ithaca, New York. And he'd certainly never have imagined that he'd join forces with one of the most talented wildlife photographers in the world and spend the better part of a decade documenting every species of bird-of-paradise in the wild. But if that young guy in Tucson could look into a crystal ball, he would see himself embarking on a journey that would be nearly as extraordinary as the birds-of-paradise themselves.

In a misty, moss-covered forest in the Arfak Mountains, Ed searches for birds-of-paradise during a respite from the daily rains.
Upper Syoubri, Bird's Head Peninsula, 3 September 2009

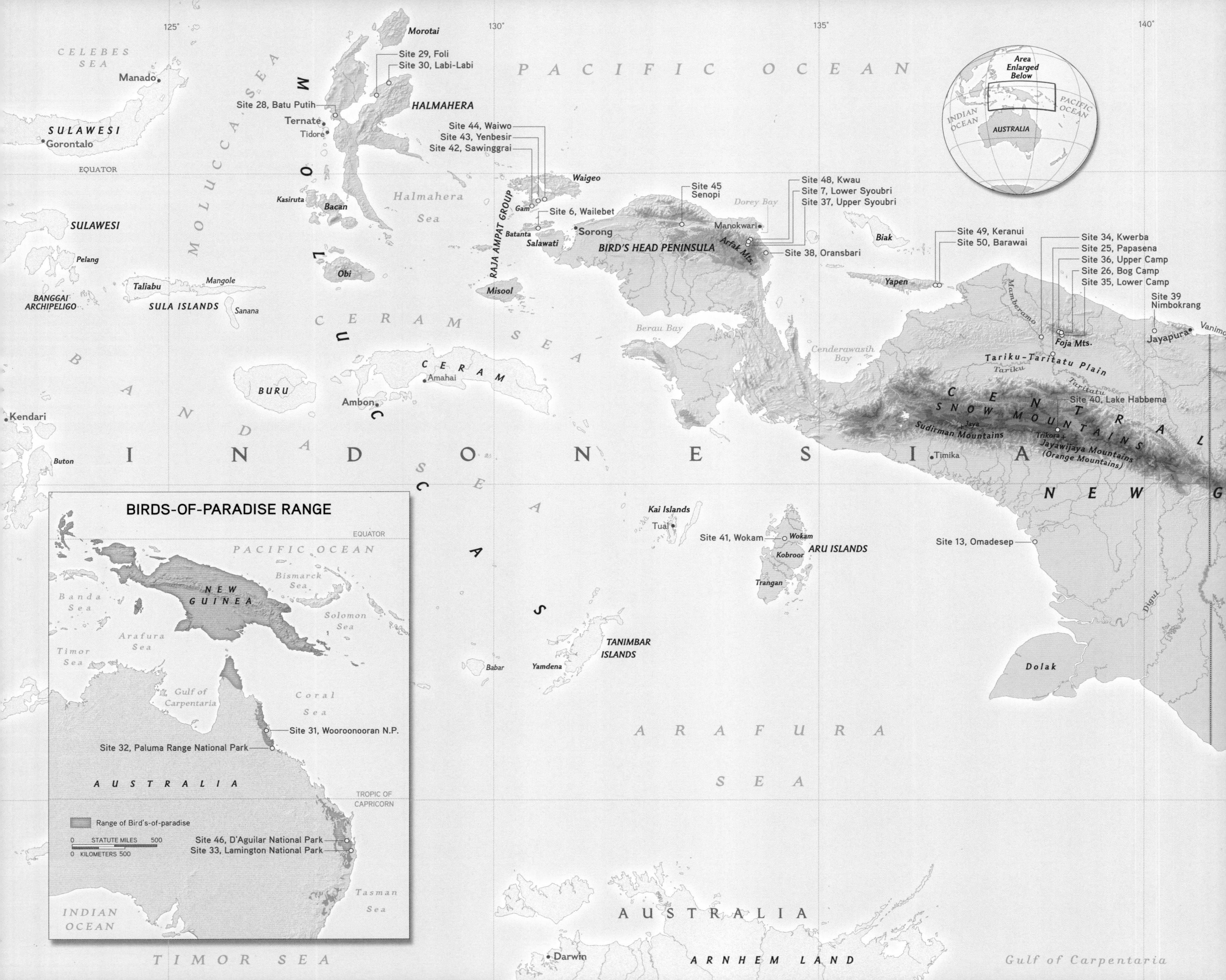

125°
130°
135°
140°
CELEBES SEA
Manado
SULAWESI
Gorontalo
EQUATOR
MOLUCCA SEA
Morotai
Site 29, Foli
Site 30, Labi-Labi
PACIFIC OCEAN
HALMAHERA
Site 28, Batu Putih
Ternate
Tidore
Site 44, Waiwo
Site 43, Yenbesir
Site 42, Sawinggrai
Waigeo
Kasiruta
Bacan
Halmahera Sea
Gam
Site 6, Wailebet
RAJA AMPAT GROUP
Batanta
Salawati
Sorong
Site 45 Senopi
Dorey Bay
Manokwari
Site 48, Kwau
Site 7, Lower Syoubri
Site 37, Upper Syoubri
BIRD'S HEAD PENINSULA
Arfak Mts.
Site 38, Oransbari
Biak
Site 49, Keranui
Site 50, Barawai
Yapen
SULAWESI
Pelang
Obi
Misool
Taliabu
Mangole
SULA ISLANDS
Sanana
BANGGAI ARCHIPELIGO
CERAM SEA
Berau Bay
Cenderawasih Bay
Mamberamo
Site 34, Kwerba
Site 25, Papasena
Site 36, Upper Camp
Site 26, Bog Camp
Site 35, Lower Camp
Site 39 Nimbokrang
Jayapura
Vanimo
Foja Mts.
Tariku-Taritatu Plain
Tariku
Taritatu
BANDA SEA
BURU
CERAM
Amahai
Ambon
MOLUCCAS
Kendari
Buton
INDONESIA
CENTRAL
SNOW MOUNTAINS
Site 40, Lake Habbema
Jaya
Sudirman Mountains
Trikora
Jayawijaya Mountains (Orange Mountains)
Timika
NEW G
Kai Islands
Tual
Site 41, Wokam
Wokam
ARU ISLANDS
Kobroor
Trangan
Site 13, Omadesep
Digul
TANIMBAR ISLANDS
Babar
Yamdena
Dolak
ARAFURA SEA
AUSTRALIA
Darwin
ARNHEM LAND
TIMOR SEA
Gulf of Carpentaria
Area Enlarged Below
INDIAN OCEAN
PACIFIC OCEAN
AUSTRALIA
BIRDS-OF-PARADISE RANGE
EQUATOR
PACIFIC OCEAN
NEW GUINEA
Bismarck Sea
Banda Sea
Solomon Sea
Arafura Sea
Timor Sea
Gulf of Carpentaria
Coral Sea
Site 31, Wooroonooran N.P.
Site 32, Paluma Range National Park
AUSTRALIA
TROPIC OF CAPRICORN
Range of Bird's-of-paradise
0 STATUTE MILES 500
0 KILOMETERS 500
Site 46, D'Aguilar National Park
Site 33, Lamington National Park
Tasman Sea
INDIAN OCEAN

BIRDS-OF-PARADISE PROJECT EXPEDITIONS AND FIELD SITES

2004 - Expedition 1
Site 1 - Upper Pass, Mount Hagen
Site 2 - Lower Pass, Mount Hagen
Site 3 - Kama, Wapenamanda area
Site 4 - Sebutuia, Fergusson Island
Site 5 - Dobu Island

2004 - Expedition 2
Site 6 - Wailebet, Batanta
Site 7 - Lower Syoubri, Arfak Mountains

2005 - Expedition 3
Site 1 - Upper Pass, Mount Hagen
Site 8 - Karawari River, Sepik region
Site 9 - Rondon Ridge, Kubor Range

2005- Expedition 4
Site 1 - Upper Pass, Mount Hagen
Site 10 - Lower Heroana, Crater Mountain
Site 11 - Koko-o, Crater Mountain
Site 12 - Upper Heroana, Crater Mountain

2006 - Expedition 5
Site 13 - Omadesep, Asmat region

2006 - Expedition 6
Site 14 - Arafundi River, Sepik region
Site 15 - Tigibi, Tari area
Site 16 - Ambua, Tari area
Site 17 - Lower Tari Gap
Site 18 - Mid Tari Gap
Site 19 - Upper Tari Gap

2006 - Expedition 7
Site 20 - Araurang, Huon Peninsula
Site 21 - Gatop, Huon Peninsula
Site 22 - Lower Satop, Huon Peninsula
Site 23 - Kotet, Huon Peninsula
Site 24 - Towet, Huon Peninsula

2007 - Expedition 8
Site 25 - Papasena, Mamberamo region
Site 26 - Bog Camp, Foja Mountains

2007 - Expedition 9
Site 21 - Gatop, Huon Peninsula
Site 27 - Upper Satop, Huon Peninsula

2008 - Expedition 10
Site 28 - Batu Putih, Halmahera
Site 29 - Foli, Halmahera
Site 30 - Labi-Labi, Halmahera

2008 - Expedition 11
Site 31 - Wooroonooran N.P., Australia
Site 32 - Paluma Range N.P., Australia
Site 33 - Lamington N.P., Australia

2008 - Expedition 12
Site 7 - Lower Syoubri, Arfak Mountains
Site 26 - Bog Camp, Foja Mountains
Site 34 - Kwerba, Mamberamo region
Site 35 - Lower Camp, Foja Mountains
Site 36 - Upper Camp, Foja Mountains

2009 - Expedition 13
Site 7 - Lower Syoubri, Arfak Mountains
Site 37 - Upper Syoubri, Arfak Mountains
Site 38 - Oransbari, Bird's Head Peninsula

2010 - Expedition 14
Site 7 - Lower Syoubri, Arfak Mountains
Site 39 - Nimbokrang, Jayapura area
Site 40 - Lake Habbema, Snow Mountains

2010 - Expedition 15
Site 41 - Wokam, Aru Islands
Site 42 - Sawinggrai, Gam
Site 43 - Yenbeser, Gam
Site 44 - Waiwo, Waigeo
Site 45 - Senopi, Bird's Head Peninsula

2010 - Expedition 16
Site 1 - Upper Pass, Mount Hagen
Site 16 - Ambua, Tari area
Site 46 - D'Aguilar N.P., Australia
Site 47 - Kiburu, Mendi area

2011 - Expedition 17
Site 7 - Lower Syoubri, Arfak Mountains
Site 48 - Kwau, Arfak Mountains
Site 49 - Keranui, Yapen
Site 50 - Barawai, Yapen

2011 - Expedition 18
Site 4 - Sebutuia, Fergusson Island
Site 51 - Sombom, Huon Peninsula
Site 15 - Tigibi, Tari area

BIRDS-OF-PARADISE
Present
Absent

ELEVATIONS WITHIN BIRDS-OF-PARADISE RANGE
4,884 meters
4,000 meters
3,000 meters
2,000 meters
1,000 meters
Sea Level

STATUTE MILES 0 100 200
KILOMETERS 0 100 200 300 400
SCALE 1 : 7,000,000
Mercator Projection

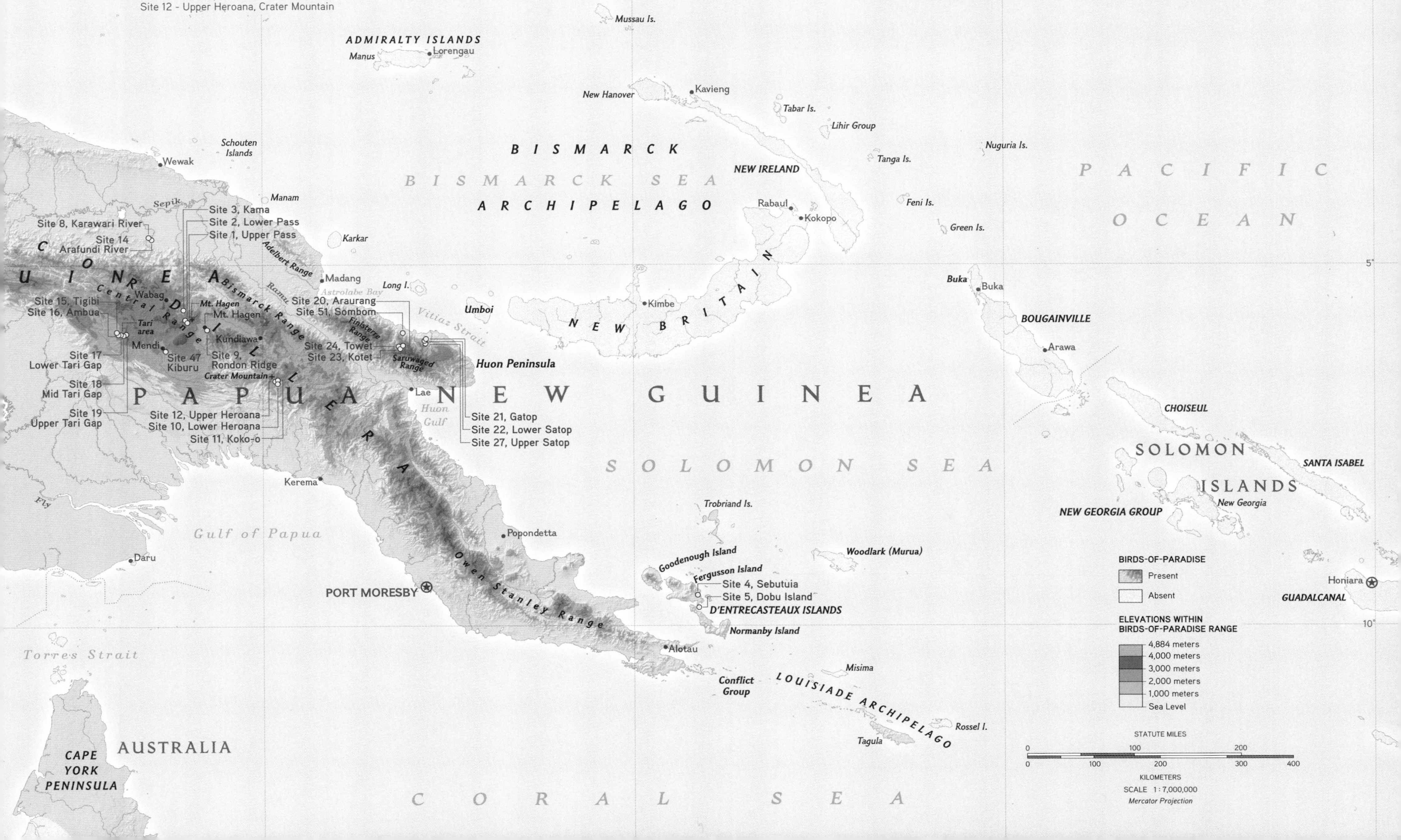

39 SPECIES

The Birds-of-Paradise

RIBBON-TAILED ASTRAPIA

WILSON'S BIRD-OF-PARADISE

CRINKLE-COLLARED MANUCODE

PALE-BILLED SICKLEBILL

TWELVE-WIRED BIRD-OF-PARADISE

KING OF SAXONY BIRD-OF-PARADISE

RAGGIANA BIRD-OF-PARADISE

WAHNES'S PAROTIA

HUON ASTRAPIA

GREATER BIRD-OF-PARADISE

BLUE BIRD-OF-PARADISE

BRONZE PAROTIA

TRUMPET MANUCODE

MAGNIFICENT RIFLEBIRD

RED BIRD-OF-PARADISE

PARADISE RIFLEBIRD

GLOSSY-MANTLED MANUCODE

STEPHANIE'S ASTRAPIA

CAROLA'S PAROTIA

ESSER BIRD-OF-PARADISE
VICTORIA'S RIFLEBIRD
BLACK-BILLED SICKLEBILL
BROWN SICKLEBILL
WESTERN PAROTIA

PLENDID ASTRAPIA
STANDARDWING BIRD-OF-PARADISE
CURL-CRESTED MANUCODE
EMPEROR BIRD-OF-PARADISE
JOBI MANUCODE

UPERB BIRD-OF-PARADISE
LONG-TAILED PARADIGALLA
KING BIRD-OF-PARADISE
PARADISE-CROW
LAWES'S PAROTIA

Most Beautiful & Most Wonderful

When seen in this attitude,
the bird of paradise really deserves its name,
and must be ranked as one of
the most beautiful and most wonderful
of living things.

—ALFRED RUSSEL WALLACE, *THE MALAY ARCHIPELAGO*, 1869

Both Alfred Russel Wallace and Charles Darwin used the same words: "most beautiful and most wonderful" in a coincidence of colorful Victorian prose *(see preceding and following quotes)*. Yet these words, like the luminaries who wrote them, are important because they symbolize different aspects of the years-long journey we have undertaken to reveal the hidden lives of the 39 species in the remarkable avian family Paradisaeidae: the birds-of-paradise.

Wallace used the phrase to describe a tree full of Greater Birds-of-Paradise in the grand splendor of courtship. He was in the Aru Islands, not far from where we would view the same species more than 150 years later. There he made a significant discovery. As he wrote in an 1857 letter posted from the Aru Islands, "I have discovered their true attitude when displaying their plumes, which I believe is quite new information; they are then so beautiful and grand."

Wallace's observations made him the first Western naturalist to understand how the distinctive bird-of-paradise plumes were actually used. He discovered that the truth was even more astonishing than anyone had imagined. By seeing adult males during courtship, Wallace observed how the radiant yellow flank-plumes become even more stunning when expanded over the backs of multiple males, all moving in exaggerated synchrony. It took an arduous journey to a remote wilderness in a far-flung part of the globe, but Wallace's observations made him a pioneer in the scientific study of the birds-of-paradise. He was also the first person to go down the exploratory path we followed during the course of our multi-year project. It is the path of discovery that comes from seeing birds-of-paradise in their natural setting, using their exceptional plumes in the context in which they evolved: during courtship display. There in the Aru Islands, Wallace was the first Western naturalist to realize that truly understanding the birds-of-paradise means seeing their behavior in the wild.

Observing birds-of-paradise in the wild and investigating their appearances and behaviors in novel ways were goals that lay at the heart of this project from the very beginning. We, like many before us, are captivated by the birds-of-paradise for their extraordinary and often bizarre brand of beauty, and we wanted to share our experience with the world through the lens of documentary photography and videography. Without seeing the birds-of-paradise interact with one another during courtship, we would never have been able to fully understand or explain them. So we followed in the footsteps of Wallace, the explorer and collector, and ventured into many little-known parts of the world in our search for the "true attitude" of the birds-of-paradise.

Our journeys took us all over the island of New Guinea, to its many satellite islands, and to the great continent of Australia below it. Like Wallace, our goal was also to amass a comprehensive collection of natural history specimens for the service of science and the general public. Wallace's specimens consisted of the traditional kind: carefully prepared and preserved museum skins for both scientific research and educational display. Ours are a more modern variety: digital photos and audio-video recordings, collecting tools that Wallace couldn't have even imagined. Yet our modern digital specimens were collected with the same attention to detail and for the same purposes of description and future study. Wallace's specimens were acquired by major natural history collections, and many of them, even 150 years later, are still usable. Ours are being preserved in a natural history collection of media at the Cornell Lab of Ornithology, and we hope that, like Wallace's, ours will have the same lasting value 150 years from now.

Likewise, just as some of Wallace's specimens were mounted and put on public display so that their marvels could be more widely shared, so too are ours being shared in the pages of this book and in several related educational forums, including the pages of *National Geographic* magazine, a documentary film, and a major museum exhibition. We believe that Wallace would recognize our labors as the continuation of the work he started, and we hope he would appreciate the progress made in illuminating a subject so dear to his heart.

The other luminary leading us down a path of discovery in our quest to survey all the birds-of-paradise is Charles Darwin. Darwin used the phrase "most beautiful and most wonderful" in the closing line of his famous book about evolution, *Origin of Species*. Those concluding words may echo Wallace's, but they refer to something entirely different. Rather than marveling at the beauty of a *particular* organism as did Wallace, Darwin's words reflect amazement for the seemingly endless diversity of forms that have evolved *among all* species. Darwin reminds us that the birds-of-paradise story is fundamentally one about the evolution of biological diversity. It's a story about the "endless forms" of size, shape, color, plumage, and behavior that have arisen among 39 closely related species. Exploring this diversity, including its causes and consequences, is an important part of our effort to reveal their secrets.

People are primarily drawn to the birds-of-paradise for two reasons: the remarkable appearances of any given species and the incredible diversity among them. We find the birds-of-paradise intriguing because they stray far from our mental image of a normal

BLUE BIRD-OF-PARADISE ~ *Paradisaea rudolphi*
Tigibi, Tari area, 13 September 2006

Some consider the Blue Bird-of-Paradise the most beautiful of all. The brilliant blue back feathers of the adult male are indeed a wonder to behold, seeming at certain angles and in certain lighting conditions to radiate from within. *(page 22)*

GREATER BIRD-OF-PARADISE ~ *Paradisaea apoda*
Wokam, Aru Islands, 18 September 2010

Head framed by a single stray plume, this adult male is a distant relative from the same population that inspired Alfred Russel Wallace to journey to the remote Aru Islands in the first place. For centuries, the Greater Bird-of-Paradise has been cherished for the ethereal beauty of its golden flank plumes, which during courtship explode in a filamentous array over the male's back. *(pages 24-25)*

bird. They remind us of mythical creatures from another age and challenge our preconceptions of nature. The birds-of-paradise seem paradoxical because they contradict the common view of nature that is, as Tennyson writes, "red in tooth and claw." We're left wondering how natural selection, with its emphasis on efficiency, could even permit creatures as overtly extravagant as the birds-of-paradise to exist at all.

Still, exceptional appearances are only part of the intrigue. We are also drawn to the birds-of-paradise because of how different they are from one another. If the family Paradisaeidae comprised 39 astonishingly beautiful, but similar-looking, species, it would hold only a fraction of the allure it does. It is the great *differences* that have evolved among the 39 species that make the birds-of-paradise so remarkable. How could a single family of birds evolve to be so different from one another? Among other avian families—finches or sparrows, for example, or even ducks or pigeons—the various species within them appear mostly similar: A duck is a duck. Yet among the birds-of-paradise, the various species are often wildly and unexpectedly different from one another. The evolution of such differences among species is even more incredible than the evolution of the extreme appearances within any one species.

The very things that average people find intriguing about the birds-of-paradise—exceptional appearances and unparalleled diversity—are the very things that scientists have pondered for generations. As with many other topics in biology, it was Charles Darwin who helped supply the answers. In addition to discovering natural selection (along with Wallace), Darwin also discovered the process of sexual selection, providing the framework for understanding how both the remarkable extremes and the unprecedented diversity among the birds-of-paradise have come to be. Darwin's evolutionary perspective therefore gives our journey to explore and understand the birds-of-paradise its scientific foundation. We might say that if Wallace inspired our physical journey to observe birds-of-paradise in their remote wilderness homes, then Darwin inspired our intellectual journey to make scientific sense of it all.

The birds-of-paradise really do deserve to be called the "most beautiful and most wonderful of living things." They are a treasure of our shared heritage on planet Earth, and while they are found only in one small corner of it, they draw the appreciation of virtually everyone, from every part of the globe. Providing the people of the world that opportunity—the chance to feel awe-inspired by the breadth and variety of the natural world—is one of the objectives of this book and the multiyear project that led to it. ■

KING BIRD-OF-PARADISE ~ *Cicinnurus regius*

Oransbari, Bird's Head Peninsula, 29 August 2009

A King Bird-of-Paradise opens its wings as if to defy its diminutive stature. The glowing crimson head and brilliant white underside show why this bird was crowned "the king" despite being one of the smallest birds-of-paradise.

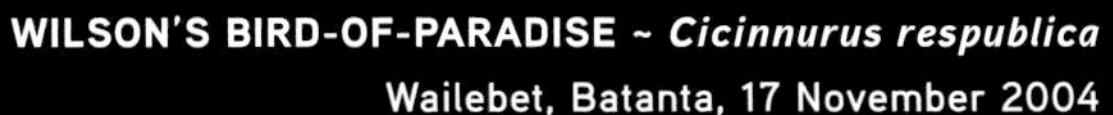

WILSON'S BIRD-OF-PARADISE ~ *Cicinnurus respublica*
Wailebet, Batanta, 17 November 2004

The Wilson's Bird-of-Paradise stands as a paragon of the "endless forms" for which the birds-of-paradise are renowned. Their extremes in size, shape, color, and behavior seem to have been literally embodied in the males of this one species.

From so simple a beginning, endless forms most beautiful and most wonderful have been, and are being, evolved.

—CHARLES DARWIN, *ORIGIN OF SPECIES*, 1859

RIBBON-TAILED ASTRAPIA ~ ***Astrapia mayeri***

Upper Pass, Mount Hagen, 1 August 2005

The beaming white tail of this adult male Ribbon-tailed Astrapia, three times longer than his body, stands in stark contrast to the leafy green background where he forages for *Schefflera* fruit. Only in a land with abundant resources and few predators could such extravagance evolve. *(left)*

BROWN SICKLEBILL ~ ***Epimachus meyeri***

Upper Pass, Mount Hagen, 12 December 2010

This Brown Sicklebill presents the quintessential example of how a female-plumaged individual could be either a female or young male (here, it is a young male). Male birds-of-paradise may take five or six years to reach their full adult plumage that contrasts distinctly with the female plumage. *(right)*

RIBBON-TAILED ASTRAPIA ~ *Astrapia mayeri*

Upper Pass, Mount Hagen, 28 July 2005

Caught in the act of shameless begging and almost gratuitous parental care, an adult female passes a single *Schefflera* fruit to this fully grown young bird, possibly a male. Not much is known about extended parental care in birds-of-paradise, but the sight of a young astrapia begging for food well beyond the age when it can fend for itself is not uncommon.

A CLOSER LOOK

The first European naturalists to examine these birds were so awestruck by their beauty and by the mystery surrounding them, they could only attribute their existence to a supernatural source.

BIRDS FROM PARADISE

HISTORY OF DISCOVERY

It was while standing on the stone ramparts of a 16th-century Portuguese fort overlooking the legendary spice-island port of Ternate that the role played by birds-of-paradise in world history became a reality for us. In the distance loomed the volcanic spire of Tidore. This is where the birds-of-paradise first became known to the Western world. The story of their discovery by Europeans and the legends surrounding their origins are as extraordinary as the birds themselves. Beyond being a wonder of the natural world, the birds-of-paradise also shaped the face of history for all humankind.

In September 1522 the Victoria, the only ship to complete Ferdinand Magellan's voyage around the world, returned to Europe after three grueling years at sea. Stowed in the cargo holds, among the many riches acquired from exotic places around the globe, were several well-preserved, wingless and legless yet unusually beautiful dead birds. These had been transported back across the seas as gifts for the King of Spain. As chronicled by crew member Antonio Pigafetta, the birds were provided by the Sultan of Bachian [now Bacan]. He had come out to visit the ship in 1521 while it was anchored off the Moluccan island of Tidore—one of the legendary Spice Islands and the ultimate destination of the grand voyage. Pigafetta writes:

> He also gave for the King of Spain two most beautiful dead birds . . . They have no wings, but instead of them long feathers of different colors, like plumes . . . They never fly, except when the wind blows. The people told us that those birds came from the terrestrial paradise, and they called them *bolon diuata,* that is to say, Birds of God.

The first European naturalists to examine these birds were so awestruck by their beauty and by the mystery surrounding them, they could only attribute their existence to a supernatural source. To the 16th-century European mind, these ethereal birds must have come from the biblical Garden of Eden. They were, as Pigafetta described, none other than birds *from* paradise.

The colorful specimens caused a minor sensation as word of their existence spread throughout Europe. In 1523, Pigafetta took the birds to Rome and presented them to Pope Clement VII as part of his account of the now famous first voyage around the world. Although the ultimate fate of these particular specimens is unknown, they were immortalized by at least two Renaissance artists: Giulio Clovio, a miniaturist who created the illuminated prayer book *The Hours of the*

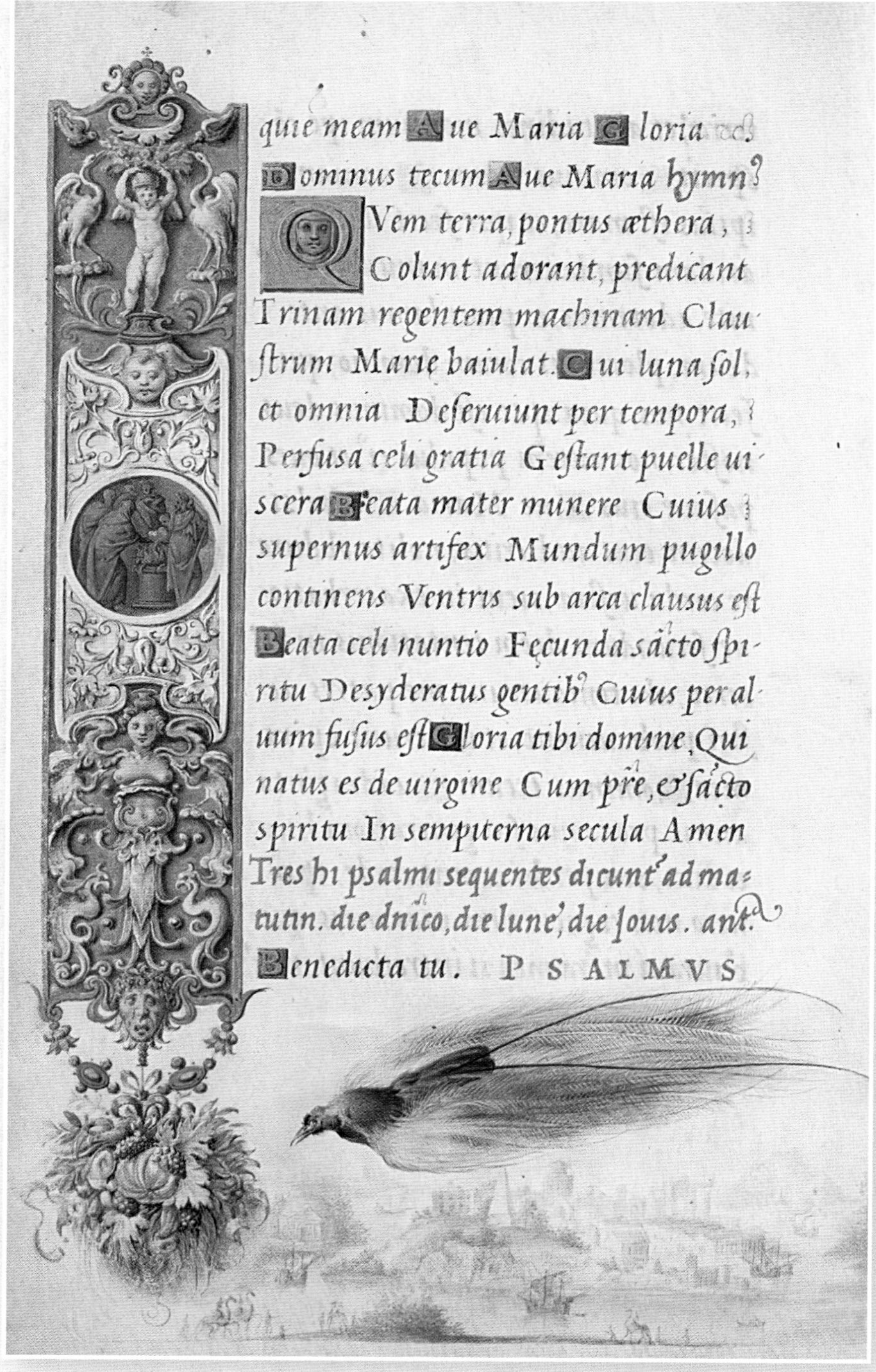
quę meam Aue Maria Gloria
Dominus tecum Aue Maria hymn?
QVem terra, pontus æthera,
Colunt adorant, predicant
Trinam regentem machinam Clau-
strum Marię baiulat. Cui luna sol,
et omnia Deseruiunt per tempora,
Perfusa celi gratia Gestant puelle ui-
scera Beata mater munere Cuius
supernus artifex Mundum pugillo
continens Ventris sub arca clausus est
Beata celi nuntio Fęcunda sacto spiritu Desyderatus gentib' Cuius per al-
uum fusus est Gloria tibi domine, Qui
natus es de uirgine Cum pře, & sacto
spiritu In sempiterna secula Amen
Tres hi psalmi sequentes dicunt' ad ma-
tutin. die dnico, die lune, die Iouis. ant.
Benedicta tu. PSALMVS

The illuminated prayer book known as *The Farnese Hours* contains one of earliest known color depictions of a bird-of-paradise. The model for this Greater Bird-of-Paradise may be one of the original wingless and legless trade skins that returned to Spain with Magellan's crew in 1522.

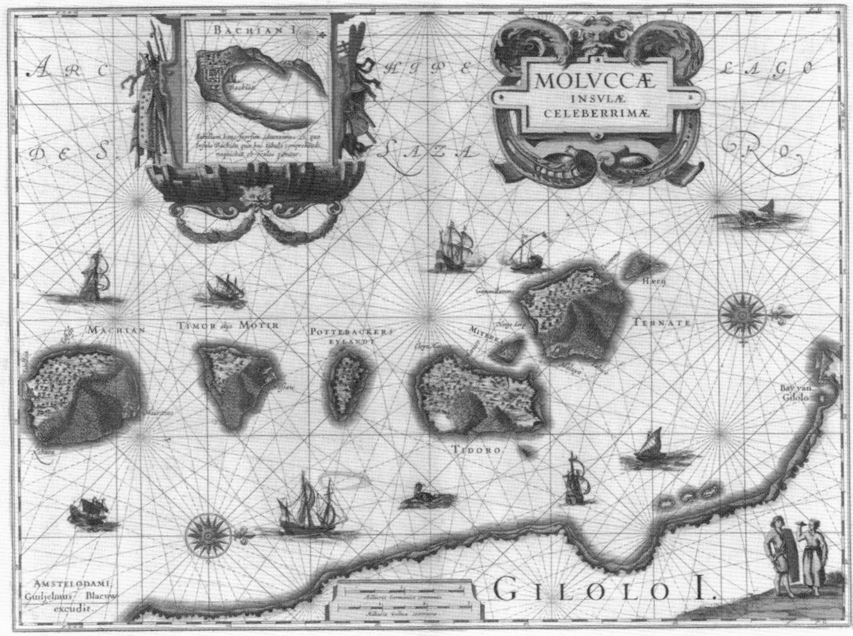

Dutch cartographer Willem Janszoon Blaeu's 1640 map of the legendary Spice Islands clearly depicts the Portuguese fort overlooking the harbor at Ternate *(shown below)* and the volcanic cone of Tidore, where the first bird-of-paradise trade skins were given to Magellan's crew as gifts for the King of Spain.

This pen-and-ink sketch by Dutch master Rembrandt van Rijn from around 1630 shows that bird-of-paradise trade skins were still a significant natural curiosity more than 100 years after their initial discovery by Westerners in 1522.

Visible looking south from a restored Portuguese fort dating to 1521 is Magellan's destination, the Spice Islands of Ternate *(foreground)* and Tidore *(volcanic island on right)*, with Halmahera (called Gilolo on old maps) in the distance. Tidore is where Magellan's crew acquired bird-of-paradise skins, the first to reach Europe. Ternate, Moluccas, 26 July 2008

Blessed Virgin Mary (also known as *The Farnese Hours*), begun in the 1530s and completed around 1546; and Giulio Romano, a student of the Renaissance master Raphael, who produced an exquisitely detailed tapestry, *The Barque of Venus,* completed around 1540. Both men are known to have been in Rome at the time of Pigafetta's papal visit, and they were so inspired by the beautiful birds from the earthly paradise that they both incorporated what appears to be the same individual specimen, a Greater Bird-of-Paradise, into their respective masterworks. These are the earliest known color depictions of any bird-of-paradise in the Western world. (A black-and-white sketch by German artist H. B. Grien, dated from 1522–25, predates them.) The appeal of illustrating "birds from paradise" would continue to inspire many generations of artists throughout the 16th and 17th centuries.

Around 1550, the great Swiss naturalist Conrad Gesner illustrated seven birds-of-paradise in the third volume (*Vogelbuch,* "Bird Book") of his four-volume masterwork *Historiae animalium.* Though it was cutting-edge natural history documentation at the time, the book nevertheless blurs the line between mythical and real creatures known to the 16th-century world. Alongside easily identified species are scientific descriptions of a phoenix, a winged lion-bodied griffin, and even a human-headed harpy. In this context, legless birds from an earthly paradise were considered entirely appropriate subject matter. Gesner's book also reported what was known about their behavior at the time. One illustration shows that a wingless and legless bird that never lands could survive by floating upright to drink dew from the clouds.

Gesner's influence was so strong that the birds-of-paradise in his *Vogelbuch* were widely copied. The works of Jan Jonston in 1657 and Ulyssis Aldrovandi in 1681 display nearly identical renditions. Dutch master Rembrandt van Rijn, known to have accumulated specimens from around the world, purportedly had bird-of-paradise skins in his private collection. He departed from previous works in two pen-and-ink sketches drawn around 1630. Yet even in Rembrandt's day, more than a hundred years after their discovery, the true nature of the birds-of-paradise remained enigmatic. Uncertain if they had legs in life or not, Rembrandt hedged his bets and drew the birds on two sheets: one sheet with and one without legs. This early attempt at scientific accuracy did not, however, influence the narrative about creatures from an earthly paradise. The myth about these wonderful birds continued to retain its hold on the popular imagination, even in the face of contrary evidence arriving on ships returning from the Far East, until well into the 18th century. —ES

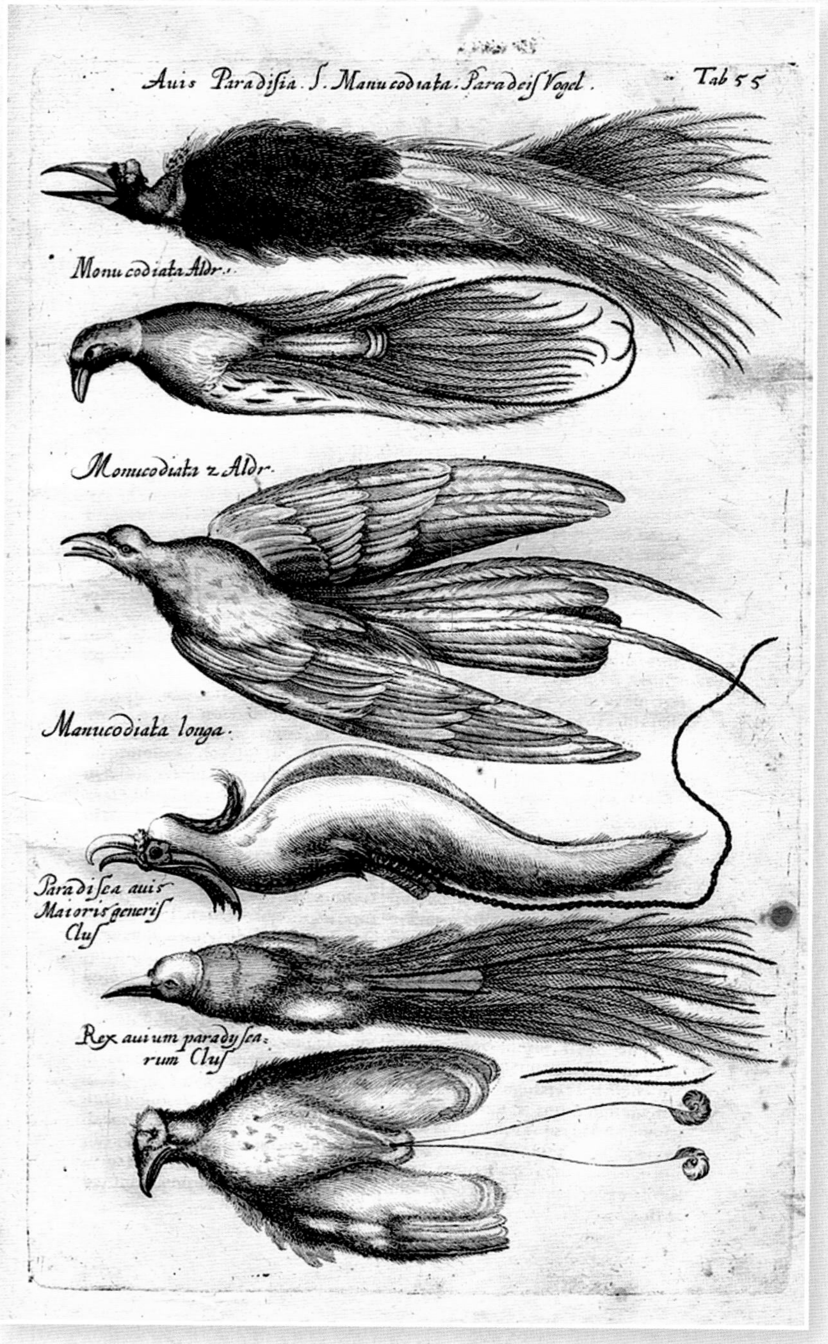

A page from Jan Jonston's 1657 book *Historiae naturalis* illustrates the entire array of birds-of-paradise known at the time. A few of the species depicted are reasonably true to life, others quite fanciful.

This map of the world from 1594 by Petrus Plancius depicts a wingless and legless cometlike bird-of-paradise as one of the iconic images of the mythical realm of "Magallanica."

[T]hese birds are never seen alive, but being dead they are found upon the Iland: they flie, as it is said alwaies into the Sunne, and keepe themselves continually in the ayre, without lighting on the earth, for they have neither feet nor wings, but onely head and body, and the most part table.

JAN HUYGEN VAN LINSCHOTEN, 1596, *ITINERARIO*

LESSER BIRD-OF-PARADISE ~ *Paradisaea minor*
Oransbari, Bird's Head Peninsula, 25 August 2009

The late afternoon sunlight catches the spectacular yellow and white flank-plumes of this adult male as he approaches the communal lek tree—a location for courtship display shared by adult males.

What Is a Bird-of-Paradise?

Birds-of-paradise are not, as some mistakenly think, merely any and every exotic-looking bird. Rather they are a group of 39 closely related species descended from an ancestor common to them and no other species. This group, called the Paradisaeidae, has the taxonomic rank of family within the order Passeriformes (passerines or perching birds). Other passerine families include the crows and jays (Corvidae), thrushes (Turdidae), starlings (Sturnidae), wrens (Troglodytidae), sparrows (Passeridae), and finches (Fringilidae). Of these, the birds-of-paradise are most closely related to the crows and jays in the family Corvidae

Like all closely related beings, birds-of-paradise have an array of uniquely shared traits. They all live in the same part of globe: the island of New Guinea, nearby islands, and northeastern Australia. All are primarily forest dwellers, most with an affinity for mountains. They are all fairly omnivorous but eat mostly fruit. All possess large, powerful feet and stout bills, which are useful for accessing difficult-to-extract fruit. The males and females of most species have very different appearances (i.e., they are sexually dimorphic). Males of most species are extremely ornate and use complex, often bizarre, display behaviors during courtship. Females of most species are brown and can sometimes be quite cryptic. All are relatively long-lived, perhaps as long as 30 years. In all the plumed species, the males play no role in parental care and mate with as many females as they can (a mating system called polygyny, which literally means "many females"). Finally, females of most species build nests from orchid tendrils and vines, lay just one egg, and rear a single chick per year.

For more than 400 years, until the mid-19th century, naturalists tended to label any unusual or beautiful crowlike bird found in the New Guinea region as a bird-of-paradise. This led to several species' being mistaken for birds-of-paradise. Examples are the giant honeyeater *(Macgregoria pulchra)* and the three species in the genus *Cnemophilus,* all of which were included in the family Paradisaeidae for over a century. Only recently did these species "fall from paradise" based on DNA evidence. Scientific understanding about what is a bird-of-paradise has changed over the decades, and as a consequence the number of species in the family Paradisaeidae has changed as science progresses. Today there is little uncertainty about the status of the "plumed" birds-of-paradise, all of which are quite exceptional indeed.

CURL-CRESTED MANUCODE ~ *Manucodia comrii*

Dobu Island, 25 September 2004

Closer in appearance to the crowlike ancestors of all the birds-of-paradise, this Curl-crested Manucode attends to the two naked hatchlings begging for food in their nest about 8 meters (26 feet) off the ground. Rearing two chicks is typical of the monogamous manucodes, but rare in the plumed species. *(left)*

LONG-TAILED PARADIGALLA ~ *Paradigalla carunculata*

Lower Syoubri, Arfak Mountains, 5 July 2010

Earthy shades of brown help to conceal this egg within the nest. Found by local naturalist Zeth Wonggor, this nest and egg are the first ever documented for this species. For most polygynous birds-of-paradise, one egg is the typical clutch size and females perform all the parental duties. *(below)*

PARADISE RIFLEBIRD ~ *Ptiloris paradiseus*
D'Aguilar National Park, Australia, 23 November 2010

Young males may look like females in their early years, but their bizarre behaviors often emerge before their plumes and reveal their true gender. Here, a subadult male practices the circular-wing pose used by adult males when courting a female. *(above)*

VICTORIA'S RIFLEBIRD ~ *Ptiloris victoriae*
Wooroonooran National Park, Australia, 5 September 2008

Young female-plumaged males may try to perform with the precision of their elders, but without specialized feather ornaments the effect is diminished—especially when compared to a fully plumed male like this one in the full splendor of his display. *(right)*

EMPEROR BIRD-OF-PARADISE ~ *Paradisaea guilielmi*
Gatop, Huon Peninsula, 7 August 2007

Communal courtship display territories, called leks, are the hallmark of most of the species in the genus *Paradisaea*. Here, in a canopy-top lek, two adult males hang upside down in the peak of display. *(preceding pages)*

MAGNIFICENT BIRD-OF-PARADISE ~ *Cicinnurus magnificus*
Lower Satop, Huon Peninsula, 15 December 2006

To establish his display site, a Magnificent Bird-of-Paradise will turn a patch of ground to bare earth and maintain its order daily. Other species use fallen logs, the sides of tree trunks, pole-like broken tree stumps, vertical vines, horizontal vines, and nearly every type of branch imaginable, demonstrating the great diversity of sites used by the 39 species.

New Guinea is indeed a paradise—for birds . . . There are no monkeys to grab the fruit, no squirrels to gnaw the nuts . . . no large mammalian carnivores either, so a male bird is not dangerously encumbered if he develops huge plumes, nor is it too risky for him to dance on the ground while displaying them.

—SIR DAVID ATTENBOROUGH, FOREWORD TO CLIFFORD FRITH AND BRUCE BEEHLER, *THE BIRDS OF PARADISE*, 1998

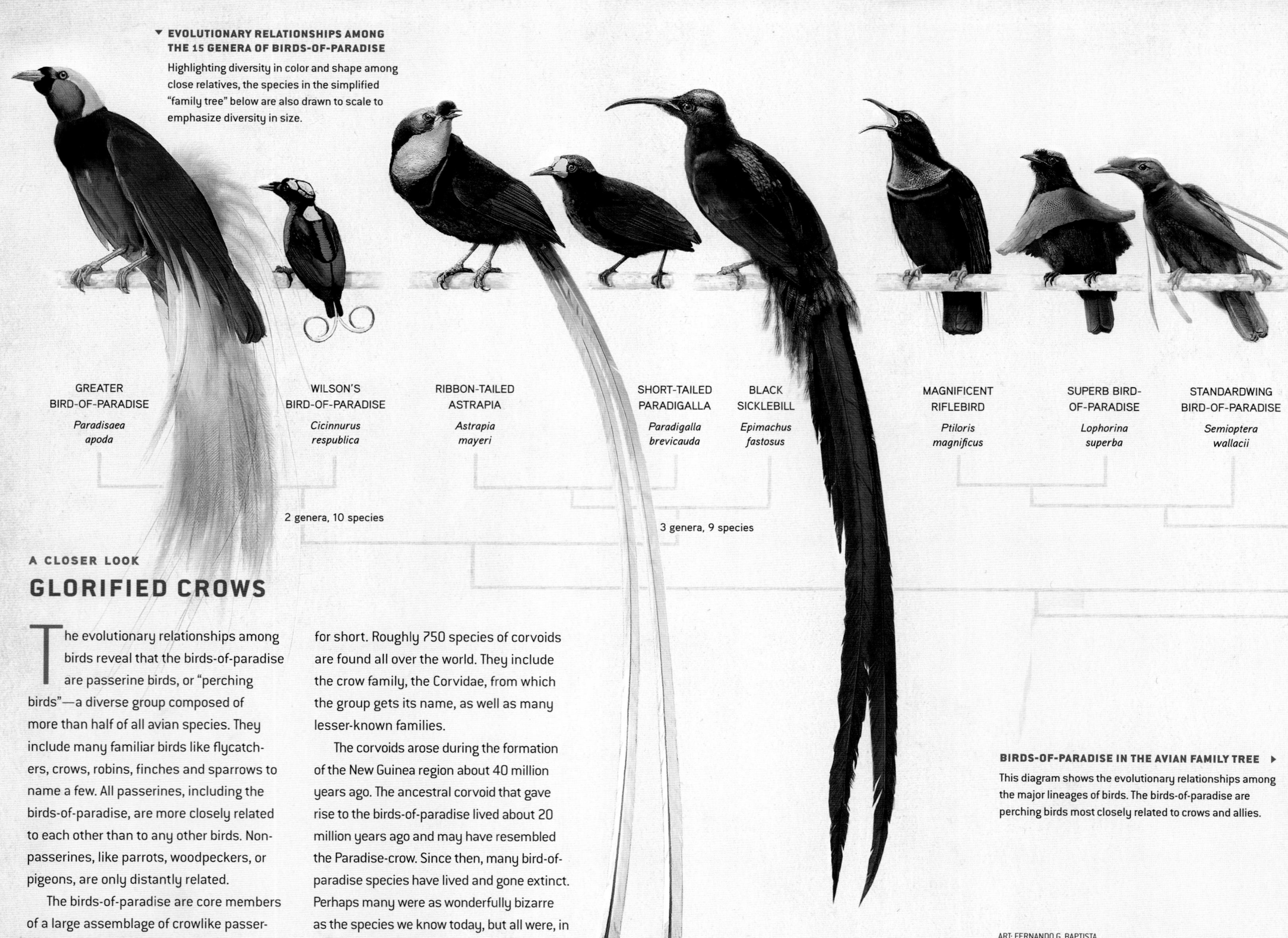

BIRDS-OF-PARADISE IN THE AVIAN FAMILY TREE ▶

This diagram shows the evolutionary relationships among the major lineages of birds. The birds-of-paradise are perching birds most closely related to crows and allies.

ART: FERNANDO G. BAPTISTA
ART RESEARCH: FANNA GEBREYESUS

A CLOSER LOOK

GLORIFIED CROWS

The evolutionary relationships among birds reveal that the birds-of-paradise are passerine birds, or "perching birds"—a diverse group composed of more than half of all avian species. They include many familiar birds like flycatchers, crows, robins, finches and sparrows to name a few. All passerines, including the birds-of-paradise, are more closely related to each other than to any other birds. Non-passerines, like parrots, woodpeckers, or pigeons, are only distantly related.

The birds-of-paradise are core members of a large assemblage of crowlike passerines known as the Corvoidea, or corvoids for short. Roughly 750 species of corvoids are found all over the world. They include the crow family, the Corvidae, from which the group gets its name, as well as many lesser-known families.

The corvoids arose during the formation of the New Guinea region about 40 million years ago. The ancestral corvoid that gave rise to the birds-of-paradise lived about 20 million years ago and may have resembled the Paradise-crow. Since then, many bird-of-paradise species have lived and gone extinct. Perhaps many were as wonderfully bizarre as the species we know today, but all were, in a sense, little more than glorified crows. —ES

Familiar to people around the world, this Rock Pigeon is drawn to scale as a relative size comparison to the birds-of-paradise above.

HUON ASTRAPIA ~ *Astrapia rothschildi*
Towet, Huon Peninsula, 11 December 2006

With one flick of the head, this female-plumaged bird will swallow whole the large *Schefflera* fruit in its bill. The many types of fruits produced by the trees and shrubs in the genus *Schefflera* are important sources of food for birds-of-paradise.

Eating the Fruits of Paradise

Birds-of-paradise, like many of their crowlike relatives, are quite omnivorous. They eat a wide range of foods, from plant-based foods to insects and other invertebrates, such as spiders. They have even been known to eat the occasional small vertebrate, like a frog or skink. However, the primary food that fuels these exuberant birds is fruit, which is abundant in the New Guinean rain forest. Birds-of-paradise may have even evolved a special type of mutually beneficial relationship with particular tree species that provide high-quality fruit. The trees evolve fruits that these powerfully billed and dexterous birds-of-paradise find desirable and accessible. The birds eat the fruit, swallow the seeds, and later regurgitate them far from the base of the parent tree, thus propagating the trees. Birds-of-paradise enjoy an abundant high-energy food source without spending much time and effort searching, and trees in turn benefit by having their offspring dispersed widely throughout the forest. It's an evolutionary win-win scenario called mutualism.

Still, the significance of fruit to the birds-of-paradise goes even deeper, for it was a fruit-based diet that permitted the evolution of the courtship extremes for which the birds-of-paradise are renowned. The ability to thrive on a wide range of fruits made finding food relatively easy. Less time and energy spent searching for food opened the "evolutionary door" to new possibilities, including the polygynous ("many female") breeding system, which fosters extreme sexual selection. For polygyny to be possible, a female must be able to spend tremendous amounts of time observing lengthy courtship displays and she must be able to provision her offspring by herself. In a polygynous system, males must be able to spend inordinate amounts of time courting females, without the competing need of hunting for food. As it turns out, the key ingredient that made polygyny possible for the birds-of-paradise is fruit.

BLUE BIRD-OF-PARADISE ~ ***Paradisaea rudolphi***

Tigibi, Tari area, 17 September 2006

With the great diversity of fruit found in the New Guinean forests comes a wide assortment of feeding strategies used to gather it by birds-of-paradise. Here an adult male adopts a pose similar to the one he uses during courtship display and hangs upside down to pluck fruit from the end of a small branch. *(left)*

PALE-BILLED SICKLEBILL ~ ***Drepanornis bruijnii***

Nimbokrang, Jayapura area, 30 June 2010

Typically in birds, an exceptionally shaped bill signifies feeding specialization. This male is using his extraordinary pale-colored bill to pry tiny fleshy seeds from a volleyball-shaped capsule high in the forest canopy. Little is known about the feeding habits of this species, but in general, birds-of-paradise consume a variety of fruits of many shapes and sizes. *(right)*

TWELVE-WIRED BIRD-OF-PARADISE ~ *Seleucidis melanoleucus*
Nimbokrang, Jayapura area, 25 June 2010

Male birds-of-paradise are highly territorial. Males like this Twelve-wired Bird-of-Paradise signal their presence to females and rivals by their calls. A male most aggressively defends his primary display site, which for this species is always a bare vertical branch or stump rising above the surrounding forest.

The males of these species are really bachelor birds who don't care a fig about domestic life.

—S. DILLON RIPLEY, *NATIONAL GEOGRAPHIC*, 1950

This was the place where Alfred Russel Wallace saw birds-of-paradise displaying in the wild, the first Western naturalist to make such observations.

IN THE FOOTSTEPS OF WALLACE

EXPLORING THE ARU ISLANDS

Aru Islander Eli Karey explained what I had to do: "Before you climb the tree in the morning, you must get some wax from your ear with your finger, and rub it on the tree trunk. You must do this so the birds will come!" Well, I thought, why shouldn't I follow the traditions of the Aru bird hunters—especially since they were being so helpful?

A few days before, it was with great anticipation that Ed and I, with our Indonesian guide Shita Prativi, had boarded a small plane in Ambon for the fabled Aru Islands. They are no longer as hard to reach as in Alfred Russel Wallace's day, but they still lie completely off the tourist track in a remote corner of Indonesia. Even before Wallace came, he knew that the Aru Islands were a source for birds-of-paradise. Because the islands were located on an ancient trade route, some of the earliest bird-of-paradise skins to reach Europe came from Aru. Both the King and Greater Birds-of-Paradise, the first two members of the family to be given scientific names by Linnaeus in 1758, originated there. This was the place where Wallace saw birds-of-paradise displaying in the wild, the first Western naturalist to make such observations. So we were making a pilgrimage of sorts to view the Greater Bird-of-Paradise where Wallace had first seen them.

Our initial reports, brought back from a scouting trip by Shita, had not been good. Hunting of birds-of-paradise seemed to still be rampant, and display sites with mature males were nearly impossible to find, even in government nature reserves. Then we had some positive news. Swiss biologist and photographer Loïc Degen, who had been visiting Aru regularly since 2001, had been teaching conservation at a local school. Because of his interest in filming the birds, a group of landowners from Wakua Village realized that their birds might be more valuable alive than hunted. They had created a sanctuary, and when we offered to pay for the permission to photograph there, they were eager to help.

That is how I found myself up in a tree, not building one of my usual platforms of boards and camouflage cloth but watching three men from the Karey clan fashion a traditional-style Aru hunting blind for me. They didn't bother with frivolities like ropes. Instead, they climbed to the canopy by stages, transferring to bigger trees twice on the way up using lengths of rattan to bend the trees together. They had insisted that the birds would not come if I built one of my blinds. The right materials and building protocols were critical, right down to the exact number of poles for the floor and wraps of

"Natives of Aru Shooting the Greater Bird-of-Paradise," from *The Malay Archipelago* (1869), suggests numerous adult males in a single lek. Such numbers in one lek are unheard of today because of hunting pressure.

Like Wallace, we traveled from Dobo to the "River" Watelai, which bisects the Aru Islands. Our field site was located just a few kilometers from Wallace's, but on the north side of Watelai Channel.

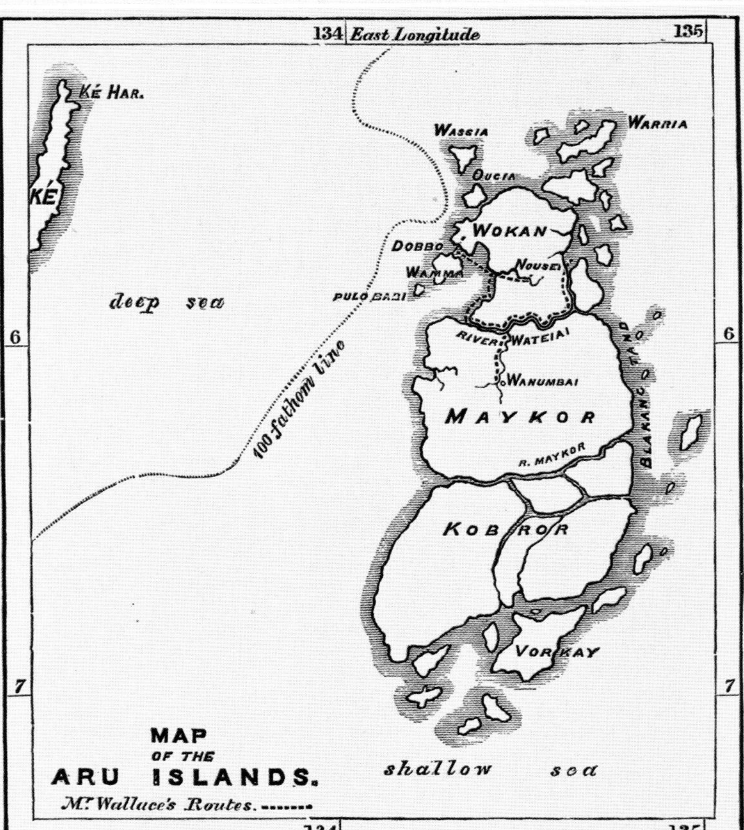

Two adult male Greater Birds-of-Paradise perch in their lek tree after a display session. Wokam, Aru Islands, 16 September 2010

Using materials from the forest, expert tree climber Eli Karey constructs a traditional Aru Islands bird-hunting blind high in a tree for Tim. Wokam, Aru Islands, 15 September 2010

vine to tie everything together. I guessed that these methods had probably been handed down across generations of Aru hunters, perhaps even from before the time of Wallace's visit. I decided it was best to go with the flow.

As I walked back into camp after my first morning up in the blind, the men came out to meet me. Had I dabbed my earwax on the tree? Had the birds come? Had I gotten any pictures? Yes to all of the above, I said. The Greater Birds-of-Paradise were spectacular, and I think I was as elated as Wallace had been upon first seeing them display. The Aru men were elated as well. All was right with the world. Hopefully now, through our project's efforts, many more people will be able to appreciate this wonder of nature.

EXPLORING HALMAHERA

One of the discoveries that most excited Alfred Russel Wallace during his eight-year sojourn in the Far East was a very strange new bird-of-paradise that he discovered in the Moluccas. Later named after him by British ornithologist G. R. Gray, it bears the Latin name *Semioptera wallacii* and is often called Wallace's Standardwing. This exciting discovery was not his only breakthrough in the sweltering, malaria-infested forests of the Moluccas. Nor was it his most important or lasting contribution. That stemmed not from individual observations but from an overarching principle that dawned on him. On one of his collecting trips in Halmahera, while he was laid up in camp by a bout of malarial fever, his ideas crystallized about the process of evolution by natural selection. In that crude field camp, he drew up the notes that became the basis of his famous letter to Charles Darwin, mailed shortly thereafter from Ternate. The rest, as they say, is history. Wallace became the co-discoverer of the unifying theory of all biological science.

After only ten days in Halmahera, Ed and I traveled to our third location. A burst of insight into evolutionary theory would have been welcome, but all we were really hoping for was to find an active lek—a courtship display area—where several male Standardwings would gather every morning. Instead, we found ourselves lying under our mosquito nets on rough floorboards in a very small garden hut we had rented, listening to the pounding rain and speculating on Wallace's fortitude. He must have been one tough dude to be able to put up with this for not just weeks but months on end. Rather than just surviving, he had reached one of the greatest insights in history. We were humbled.

We got a taste of some of Wallace's other travails during this expedition as well. He had notoriously poor luck in small boats, and the local craft we sailed in along the Halmahera coast managed to break a propeller and have a motor breakdown in the same day. We drifted for hours while repairs were made. Fortunately, the sea was flat calm and we were in no danger beyond having our backsides go numb. Yet our predicament could have become serious had a squall come up. We were much too heavily loaded to handle rough seas.

We did eventually locate an active lek of the Standardwing Bird-of-Paradise and completed our mission to document this bird—a real outlier in terms of bird-of-paradise distribution as well as behavior and appearance. Isolated for millions of years on these Moluccan islands, the Standardwing clearly had a lot of time to diverge from its cousins and develop some totally unique features like the two flaglike feathers, or standards, that emerge from the top of each wing—yet another beautiful example of the wonders of sexual selection. —TL

A male Standardwing Bird-of-Paradise raises his "standards" in display. Batu Putih, Halmahera, 16 July 2008

Male and female Standardwings were illustrated in Wallace's *The Malay Archipelago* (1869).

I now saw that I had got a great prize, no less than a completely new form of the Bird of Paradise.

ALFRED RUSSEL WALLACE,
THE MALAY ARCHIPELAGO, 1869

As dawn breaks over Foli, on Halmahera, Ed boards a local boat we used to travel along the coast. With all our gear stowed under the blue tarp, the boat was heavily loaded. Foli, Halmahera, 19 July 2008

Rugged Paradise

I look with intense interest on those rugged mountains,
retreating ridge behind ridge into the interior,
where the foot of civilized man had never trod.
There was the country of the cassowary
and the tree-kangaroo, and those dark forests
produced the most extraordinary and
the most beautiful of the feathered inhabitants
of the earth—the varied species of birds of paradise.

—ALFRED RUSSELL WALLACE, *THE MALAY ARCHIPELAGO*, 1869

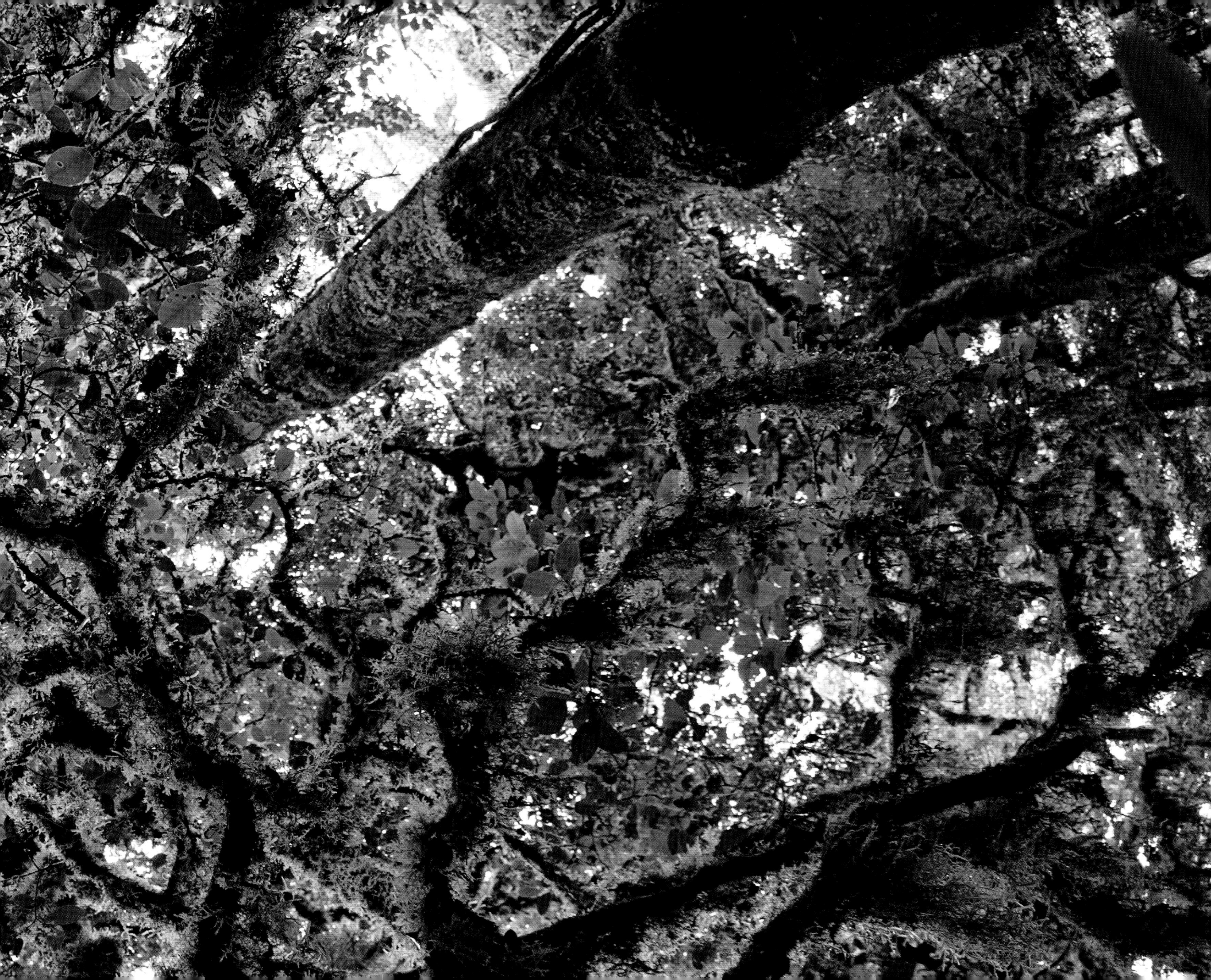

The birds-of-paradise are exceptional not only in appearance and behavior; that they evolved at all is exceptional in itself. To some extent, they are possible only because of the uniqueness of the particular landscape they inhabit. The landscape that made the birds-of-paradise possible is complex and full of contrasts. The great island of New Guinea is at once geologically young and ancient. The oldest sections are pieces of the primeval supercontinent of Gondwana. The younger sections are quite recent by geologic standards—they were borne out of the sea volcanically or lifted skyward by dynamic tectonic forces. New Guinea is also quite big—the second largest island in the world (after Greenland)—and has a tropical climate that supports tremendous biodiversity. Although it represents less than one half of 1 percent of the Earth's surface, New Guinea may harbor as much as 10 percent of the Earth's species. Since it is relatively isolated from other major landmasses, its geologic history has followed a path that created a particular haven for birds.

As in Australia, the fauna of New Guinea is, for the most part, found nowhere else on the planet. Unlike several of the large tropical islands to the west—Borneo, for example—New Guinea has never been connected by a land bridge to Asia. Its animals (and to a lesser extent its plants), like Australia's, have originated and diversified there without competition from the many lineages that have come to dominate other parts of the world. This biogeographic factor clearly marks the mammals of the region, where marsupials (like kangaroos) dominate and placental mammals (like deer) have not had the opportunity to proliferate. In New Guinea, no predatory cats or civets hunt the birds. No primates compete with birds for access to fruit. No squirrels contend with them for access to nuts and seeds. Thus in many ways, New Guinea has long been an avian paradise. In this land, many types of birds, not just birds-of-paradise, have evolved incredible diversity. Even groups of birds that have become widespread throughout much of the world, such as pigeons, kingfishers, parrots, and the crowlike corvoid relatives of the birds-of-paradise, have centers of evolutionary diversity in the New Guinea region.

The ecological similarities between New Guinea and Australia are no coincidence. A large part of southern New Guinea is actually a part of Australia, or to be more precise, part of the Australian continental shelf. The two landmasses have been connected as one larger landmass at different times in geologic history when the shallow seas that separate them receded—for example, during glacial periods when global sea levels were lower. During these periods when a land bridge connected them, plants and animals were exchanged in both directions.

By contrast, New Guinea has never been connected to any of the islands of the Malay Archipelago to the west, separated from it by deep-water trenches. As Alfred Russel Wallace famously pointed out, a stark demarcation exists between New Guinea and these islands. To the west, the fauna is largely Asian in origin. To the east, it is largely Australian (or Australo-New Guinean). These lines are not hard and fast, of course. Many plants, unlike most land animals, can disperse across open oceans without a land bridge, and so the flora of New Guinea is more of a composite than the animal life is. In addition, certain animal species, including birds that can travel many miles over open ocean, have colonized New Guinea from the Asian side (e.g., hornbills). But for the most part, the demarcation is profound, and birds-of-paradise fall entirely on the eastern side of the line. They have originated and diversified there and cannot be found at all even across certain narrow straits on islands separated by deep ocean.

Another crucial influence on New Guinea's ecological development is the parts of its landmass that never belonged to the Australian continent—or any continent, for that matter. These regions have been created in more recent geologic times. Virtually the entire northern half of this large island, as well as many nearby islands, was produced by the extreme geologic dynamism that makes this region so geographically complex. The massive spine of mountains that spans the interior from east to west is one of the most prominent outcomes of the churning forces below. These mountains have risen to the clouds because of the buckling uplift caused by the two massive tectonic plates that are colliding below the New Guinean landmass: the Indo-Australian plate and the Pacific plate. Here mountains grow at rates as fast as any on Earth.

This clash of tectonic plates has produced another effect, as well. Some parts of the central ranges and most of the mountains along the north coast spring from the mountain-building volcanism that occurs along tectonic plate boundaries. Mountainous islands that were once part of a volcanic island arc system were transported, as if riding a geologic conveyer belt, along the plates' boundaries until they collided with the New Guinean landmass. They then became fused—in geologic terms, accreted—to the landmass itself.

Thus the New Guinea we know today is a complex composite of the older Australian continental shelf, mountain ranges that were built by

Upper Pass, Mount Hagen, 29 August 2004

The first rays of sunlight pierce the blue-hued stillness of early morning in the New Guinea highlands. Somewhere out in that forested wilderness, male birds-of-paradise are starting to display, but finding them is the challenge. *(page 62)*

Bog Camp, Foja Mountains, 18 November 2008

Only when you pause to look up into the emerald green vastness of the forest canopy do you start to appreciate that the living soul of the New Guinean cloud forest is the network of moss-covered tree trunks and branches, home to the many wondrous creatures living in this rain-soaked habitat. *(pages 64-65)*

the uplift forces of colliding plates, and the relatively recent island arc fragments that have become attached to the main island at various times throughout geologic history.

All of these forces have produced a landscape of great ecological complexity and a diversity of habitats for New Guinea's plants and animals, including the birds-of-paradise. Once created, whether by uplift or tectonic fusion, New Guinea's mountains, and the valleys that separate and isolate them, became powerful engines driving the evolution of biodiversity.

New Guinea's geologic dynamism, especially mountain building, has been essential for bird-of-paradise evolution. Approximately 70 percent of bird-of-paradise species are considered montane, with substantial parts of their ranges found above 500 meters (1,650 feet). About 30 percent of them are considered high-montane, or cloud forest, species, living at elevations between 1,500 and 3,500 meters (5,000–11,500 feet) above sea level. Many of those montane species are restricted to specific mountain ranges, some of which were once the offshore islands that became fused to the mainland. These "islands" of montane forest are separated from other montane forests by "inland seas" of lowland forest.

New Guinea's mountains also have a major impact on climate. Parts of southern New Guinea are considerably more arid, at least seasonally, than most of the rest of the island, which is wet (and seasonally wetter) throughout the entire year. Birds-of-paradise are almost exclusively forest dwellers, and they don't live in grasslands or seasonally dry forests. But they do inhabit New Guinea's many coastal forests, including mangrove forests. Several species, like the Twelve-wired Bird-of-Paradise, have evolved to thrive in swamp forests, which are fairly extensive in coastal regions and the vast river basins of the Sepik and Mamberamo Rivers in the north and the Fly River in the south. Many birds-of-paradise also inhabit rain forests in lowland areas that are found around the entire island.

From the subalpine regions in the clouds to the mangroves that crowd the coast, the story of the birds-of-paradise is highly interwoven with the many habitats in which they are found. The complexity of the landscape and its ability to foster isolation have enabled them to evolve their striking differences in appearance and behavior. This rugged paradise has, as Wallace said, "produced the most extraordinary and most beautiful of the feathered inhabitants of the earth." ■

BLUE BIRD-OF-PARADISE ~ *Paradisaea rudolphi*

Upper Herowana, Crater Mountain, 3 October 2005

Calling from an exposed snag on a hillside above his courtship display area, this adult male has a commanding view of the steeply sloping ridge system in which his territory is nestled. The sheer inaccessibility of their wilderness homes is one reason that birds-of-paradise remain so poorly known even today.

MAGNIFICENT RIFLEBIRD ~ *Ptiloris magnificus*

Oransbari, Bird's Head Peninsula, 23 August 2009

The metallic blue feathers of this bird's oddly twisted neck reflect the soft hues of the lowland rain forest understory. Although most birds-of-paradise spend time high in the canopy, many also use the forest understory for courtship display or insect foraging.

Few if any groups of birds set a naturalist's imagination on fire as do the birds-of-paradise. Their very names suggest our sense of their ethereal and unearthly mystery, while their remote homes in New Guinea place them outside of the personal experience of all but a few intrepid ornithologists.

—PAUL A. JOHNSGARD, *ARENA BIRDS*, 1994

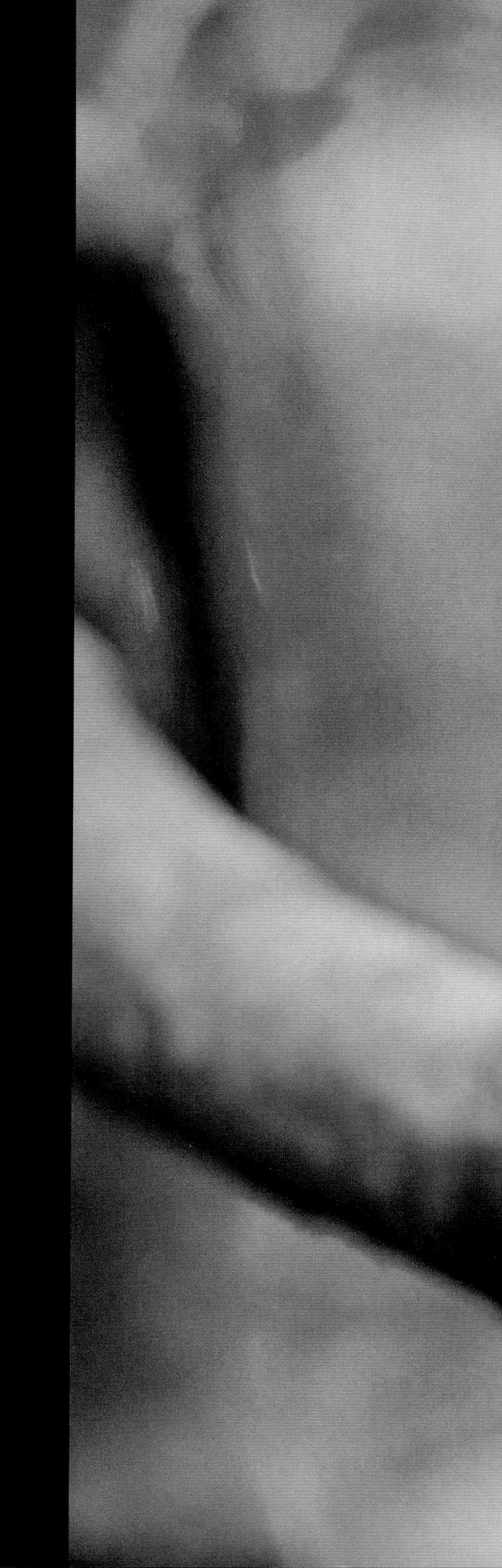

Koko-o, Crater Mountain, 30 September 2005

Small, fast-flowing, boulder-filled streams like this one, called the "water Koko-o" by the locals, are a common feature of every montane forest habitat throughout New Guinea. *(left)*

Sombom, Huon Peninsula, 11 November 2011

In most of New Guinea's mountainous terrain, a narrow muddy footpath along a ridge is the primary means of getting from one place to another. Here, Ed walks among a grove of tall narrow trees heavily laden with epiphytic growth while searching for the terrestrial display courts of Wahnes's Parotia. *(right)*

Lake Habbema, Snow Mountains, 19 June 2010

Only the nearest shoreline can be made out through the blanket of dense fog covering this spectacular high-mountain lake, which is surrounded by forests that support the gloriously plumed Splendid Astrapia. Looming large on the cloudless horizon is Trikora peak, which at 4,750 meters (15,580 feet) is the third highest peak in all of Australasia.

A CLOSER LOOK

The most extraordinary similarity between the experiences of a 21st-century and a Victorian explorer of New Guinea is the pristine state of so many of the forests. . . . The songs and calls of birds and insects fill your ears incessantly.

After a long day in the very wet and surprisingly chilly mountain forests of New Guinea, a chance to dry out and have a hot meal was always welcome.
Sombom, Huon Peninsula, 16 October 2011

LIFE IN THE FIELD

CHALLENGES OF EXPLORATION

Rain pelts the tarp-covered roof of the small hut our local assistants have fashioned from sticks, small trees, and lengths of vine used to tie the construction together. As is the custom here in the mountains of the Huon Peninsula, Tim and I sit on a bed of cut ferns with our damp feet nearly touching glowing embers. The fire "ring," which is more of a rectangle, forms the centerpiece of the hut, built in the same style as their village homes. Fire is a necessity of life in these cool, wet mountain forests, and starting the fire was the first thing the locals did upon arrival. During our weeks here, it would never be allowed to burn out.

Sitting across the fire from us, our field assistants sip the tea and coffee we provided from town. Along with the salt, sugar, soap, and protein that comes from a tin can rather than from a long day of hunting, they are the luxuries of the job helping us document the birds-of-paradise living in their forest. Draped all around us are shirts, raincoats, and various types of camera coverings we're trying to get dry. Along the edges of the shelf used to dry firewood over the fire (even wood has to be dried before use in a rain forest) hang a half-dozen pairs of damp socks accumulated over several soggy days.

As one assistant gathers plates from our meal of rice with instant noodles and edible leaves harvested from the forest, another asks us what our favorite sports are. Tim tells them he loves to ski. The four men press toward the fire, rapt with attention as Tim describes snow—a foreign substance none of them has ever seen. He relates how he attaches wide, flat sticks to his feet and zips down a snow-covered mountain faster than a cassowary can run! Tim might as well have spun a tale of growing wings and flying to the moon. The locals in turn tell us they like football (soccer) and sometimes play intervillage tournaments on holidays. Our lives are so very different, but sitting by the fire telling stories is a universal human activity.

Early New Guinea explorers, like Alfred Russel Wallace, did not possess today's expedition equipment. They brought no plastic ponchos, waterproof boxes, quick-dry socks, LED headlamps, GPS units, satellite phones, or slickly packaged energy bars. But despite these modern conveniences, it's surprising how little has changed for the 21st-century explorer in New Guinea. Sleeping on the floor of a village hut perched on stilts over a patch of muddy ground is as common today as it was in Wallace's day—as is the effort required to get a decent night's sleep

This illustration from *The Malay Archipelago* depicts a hut Wallace used as his base for several weeks on Gam Island. We often found ourselves in similar circumstances.

amid the raucous squeals of the ubiquitous pigs living underneath our heads. Whether in a forest camp or village, meals are still cooked over an open fire using hand-cut wood gathered from the forest. Food staples such as sweet potatoes, leafy greens, and bananas are still obtained from local women who harvest crops on a daily basis. From their hand-tilled gardens, the produce is carried home in large handwoven net bags hung from their foreheads and slung over their backs.

Even with the modern fabrics and waterproof gear, water remains one of the biggest enemies to success. Rains can keep you camp-bound for days on end. Precious equipment, especially electronic and photographic gear, can sprout mold if not dried properly. The same fungal-induced foot rot that hobbled Wallace for days, sometimes weeks, still awaits the explorer who thinks he can hike for eight hours in and out of shin-deep water, unable to change the expensive socks that the salesperson back home swore would keep his feet dry in any environment.

Even armed with our modern knowledge of the malarial life cycle, extra precautions taken to avoid contact with night-flying mosquitoes, and prophylactic drugs circulating in our bloodstreams, malaria continues to be a risk today as it was 150 years ago. Even relatively basic health issues can be extremely serious because doctors are few and far between. It can take days to reach a town with medical supplies of any sort. And as for major health emergencies? For the most part, we've been extremely fortunate during the project, but a truly life-threatening condition requires using a satellite phone to request a helicopter evacuation to reach a hospital. I learned that lesson the hard way when I came down with appendicitis while camped on the side of a remote mountain more than ten hours on foot from the nearest village—which didn't have a doctor anyway. Death was at the tent-fly, but I survived through a fortuitous combination of strong antibiotics, fast-acting global communications, and a helicopter evacuation that would have astounded Wallace.

The most extraordinary similarity, however, between a 21st-century and a Victorian explorer of New Guinea is the pristine state of so many of the forests. No roads and few footpaths transect them, and cell phones, radios, and televisions don't work out there. The night sky, on the rare evening when the clouds part, is unpolluted by light and provides a glittering array of stars, as the nearest town is often more than 100 miles away. The natural sounds of the forest are undisturbed by human voices and machinery, allowing the songs and calls of birds and insects to fill your ears. In this regard, the modern-day explorer can be transported to an earlier time that is becoming increasingly difficult to experience anywhere else. That's why life in the field is a key part of what makes documenting the birds-of-paradise in the wild so rewarding. —ES

Most of our field camps could be reached only on foot, and we hired many local people to help carry in our gear *(above)*. We found that local knowledge about birds varied considerably. In some places, like here in the Arfak Mountains, skilled local guides such as Zeth Wonggor *(at right, below)* greatly assisted us in finding display sites. Arfak Mountains, 11 August 2009

Tim repacks at Bog Camp, Foja Mountains. Our standard approach for rain forest camping is a small sleeping tent beneath a larger tarp to create a protected area for gear.
Foja Mountains, 24 June 2007

Gam, 3 October 2010

Isolated from the New Guinea mainland by the turquoise blue waters of the most biologically diverse coral-reef ecosystem in the world, the island paradise in the foreground is home to the spectacular Red Bird-of-Paradise—yet this species is unknown in islands just across the narrow strait.

Islands of Isolation—Coral Islands

The term "bird-of-paradise" conjures up images of a tropical paradise: blue water, palm trees, and coral-fringed beaches. We'll forget for the moment that most of the 39 species of birds-of-paradise are found in the colder, cloudier, and wetter mountain forests. New Guinea's coral-fringed islands do, in fact, play a significant role in bird-of-paradise evolution.

Around New Guinea, the world's largest tropical island, sit many smaller satellite islands that are havens for some of the most important products of biodiversity on Earth. As Darwin observed of birds in the Galapagos Islands, off South America, isolation disrupts the ability of populations to freely interbreed. As a result, animals living on islands often diverge on unique evolutionary trajectories that separate them in form and function from their geographically distant relatives. This is what has happened for six island-dwelling birds-of-paradise. Because they were nonmigratory forest species, the ancestors of the island birds didn't disperse over the open ocean, but remained completely isolated in their coral island homes.

To fully appreciate how this process occurred, it's important to remember that New Guinea sits atop one of the most complex geological hot spots on the globe. It forms part of the Pacific "Ring of Fire." Many tropical island paradises are (or were) fiery volcanoes, merely cloaked with the thin veil of paradise. These volcanic lands rise from the ocean along the boundaries where the crustal plates covering the surface of the Earth connect like pieces of a jigsaw puzzle. The Australian and Pacific plates that abut along the north coast of New Guinea create one hotbed of this activity. Here, volcanic islands emerged from the sea, got pushed along the plate boundaries, and some became fused to New Guinea's northern edge. Others merely "docked" for a geologic instant, only to break off and continue their slow migration. During those brief periods of attachment, the ancestors of the now isolated birds-of-paradise became established in their new homes. As these geologic movements continued, the ancestors of the six island species were swept along for the ride, left to evolve on their own trajectories in rarefied isolation.

WILSON'S BIRD-OF-PARADISE ~ *Cicinnurus respublica*
Wailebet, Batanta, 20 November 2004

The Wilson's Bird-of-Paradise is found only in the forests of two islands, Waigeo and Batanta, just west of mainland New Guinea. The males of this species clear and maintain the terrestrial courts they use for courtship display. Here, a male displays from a sapling by waving his metallic-blue wirelike tail feathers from side to side while a female watches intently from above. *(left)*

WILSON'S BIRD-OF-PARADISE ~ *Cicinnurus respublica*
Waiwo, Waigeo, 2 October 2010

Evolving in isolation on an independent trajectory from its closest relative on New Guinea's mainland, both males and females of this species have an unusual skullcap of bare blue skin subdivided by a narrow web of tiny feathers. *(right)*

STANDARDWING BIRD-OF-PARADISE ~ *Semioptera wallacii*
Labi-Labi, Halmahera, 24 July 2008

Found only on Halmahera and several satellite islands nearby, this species has the distinction of being the plumed bird-of-paradise with the most western distribution. With the odd elongated white "standard" feathers flaring from each wing, the plumage of the adult male *(shown here)* is nearly as peculiar as the species' distribution.

GOLDIE'S BIRD-OF-PARADISE ~ *Paradisaea decora*
Sebutuia, Fergusson Island, 22 September 2004

Calling loudly in a cacophonous courtship duet, these two males might seem to be cooperating to put on such a show, but they are also competing for access to the same females. Another example of a remarkable island endemic, this species is found only on two islands: Normanby and Fergusson, off the eastern tip of New Guinea. *(following pages)*

How . . . the paradise birds ever reached the Moluccas is one of those mysteries of bird distribution in the East Indian islands which makes study of this area so fascinating.

—S. DILLON RIPLEY, *NATIONAL GEOGRAPHIC*, 1950

A CLOSER LOOK

Wallace didn't know what we know now: that the islands of Batanta and Waigeo lie on a chunk of the Earth's crust that has been moving along the north coast of New Guinea for millions of years.

IN THE FOOTSTEPS OF WALLACE

EXPLORING THE RAJA AMPAT ISLANDS

It is fitting that some of the most beautiful islands in the world also have some of the most beautiful birds inhabiting them. The strikingly rugged, uplifted islands in the Raja Ampat region, surrounded by ultramarine waters harboring the world's richest coral reefs, are home to two species found nowhere else—the Red and Wilson's Birds-of-Paradise. This remote group of islands, near the westernmost tip of New Guinea, became a destination for Wallace and for our explorations as well.

Wallace did not travel in remote regions like this unassisted. He worked with a network of contacts, he hired assistants, and he added landowners and local guides to his team once at a location. Ed and I had to do exactly the same to pull off our expeditions. During trips in Papua New Guinea—the eastern half of the bird-of-paradise region—we worked with various researchers, missionaries, conservationists, and eco-lodges to piece together our fieldwork. Here in Papua (formerly known as Irian Jaya), the Indonesian half of bird-of-paradise land, we relied entirely on Kris Tindige and his wife and partner, Shita Prativi, to help organize all our expeditions, including our trips to the Raja Ampat islands.

Just as Wallace had done, we had to hire boats to reach the islands. We had the advantage of outboard motors, however, and so the journey took considerably less time. In retracing parts of Wallace's route, we arrived at Waigeo through the passage between that island and Gam (called Gemien by him). We reveled in the same unparalleled scenery he wrote about and landed at the village he called Bessir, now called Yenbeser, where Wallace lived many weeks while collecting Red Birds-of-Paradise. A man fished from a small dugout canoe with outriggers, a timeless scene I am certain is identical to one Wallace witnessed—except now the man occasionally pulled out his cell phone to answer a text message.

Kris got us off to a great start on our first trip with him to the island of Batanta in 2004. Unbeknownst to Wallace, nearby Batanta also harbored Red and Wilson's Birds-of-Paradise, and Kris had located an area with both species in the same vicinity. He sent a crew ahead to rig a basic camp of tarps over pole frames for us to sleep under, to gather firewood for cooking, and to dig a latrine. In other words, we had all the comforts of home and could devote our energy to fieldwork. As usual, this soon became all consuming, as we rose long before dawn to hike to display sites

Wallace based himself here, labeled as Bessir on Gemien Island on his map, while searching for the Red Bird-of-Paradise in 1860. Yenbeser, 10 May 2007

[W]e emerged into what seemed a lake but which was in fact a deep gulf . . . studded along its shores with numbers of rocky islets. . . . Every islet was covered with strange-looking shrubs and trees, and was generally crowned by the lofty and elegant palms, which also studded the ridges of the mountainous shores, forming one of the most singular and picturesque landscapes I have ever seen.

ALFRED RUSSEL WALLACE, *THE MALAY ARCHIPELAGO*, 1869

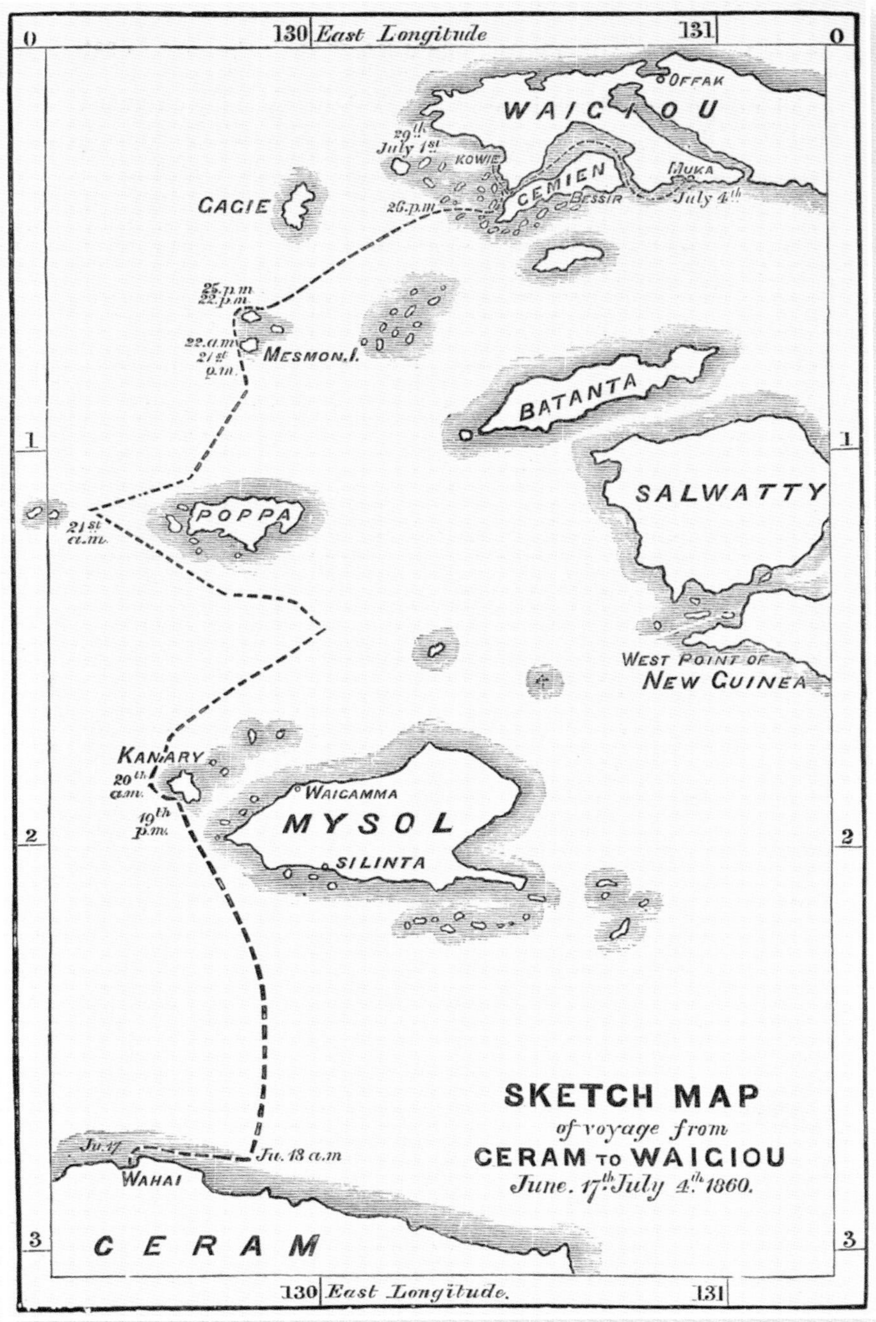

The map shows Wallace's route to Waigeo, including his passage through the channel between Waigeo and Gemien (called Gam today).

This photograph, looking to the southwest, shows the very passage taken by Wallace between Waigeo *(right)* and Gam, which he called "one of the most singular and picturesque landscapes" he had ever seen.
Gam and Waigeo, Raja Ampat Group, 3 October 2010

of Wilson's Bird-of-Paradise and put in 12- to 14-hour days in the field.

While I continued to work on Wilson's, Ed and Kris located an impressive display lek of the Red Bird-of-Paradise. Wilson's males clear courts on the ground for display, but Red Birds-of-Paradise display high in the canopy, and the tree Kris and Ed found was one of the largest in the forest. Neighboring trees were not tall enough to serve as a blind platform, but this tree was so large, I decided to climb it and see what I could do. But first I had to get up it. My bow was not powerful enough to send an arrow trailing a line over the upper branches, so I had to climb the tree in stages, targeting a lower branch and working my way up. When I finally reached the level of the display branches, I saw that the tree's crown was so large that I could build a blind on one side and still be far enough away to view the area the birds used without disturbing them. I measured the height with my rope—50 meters (165 feet)—the highest blind I would ever build for the project.

We were not far from the coast, and from my high position in the tree, I gazed out over the strait to the neighboring island of Salawati, just a few kilometers away. Amazingly, this narrow water gap between Salawati and Batanta is a major barrier to bird-of-paradise dispersal. Even though fruit pigeons and hornbills may cross the strait on a daily basis, birds-of-paradise reside in the forest and never fly over open water. For this reason, over the course of millennia separate populations have been maintained on each side of the strait. Salawati is very close to the big island of New Guinea and must have been connected at some recent point in time, for it harbors King, Lesser, and Twelve-wired Birds-of-Paradise, species typical in the New Guinea lowlands. Batanta, in contrast, has two very distinct species: Red and Wilson's Birds-of-Paradise.

Wallace didn't know what we know now: that the islands of Batanta and Waigeo lie on a chunk of the Earth's crust that has been moving along the north coast of New Guinea for millions of years. At some point, they must have bumped up against the big island and picked up some birds-of-paradise. Subsequently pulling away, time and selection led to the origin of distinct species. Though Wallace didn't know about these geological explanations, he saw the variations in species across islands throughout his collecting expeditions, and he understood that related species had diverged due to isolation. Just as the Galápagos helped Darwin gain his insights into natural selection, so the Indonesian archipelago was Wallace's source of inspiration.

Unfortunately, Kris, who also fully appreciated Indonesia's biodiversity, fell victim to cancer at only 41 and died in 2007. Since then, Shita has passionately carried on his dedication to the people of Papua and their unique birdlife. She has been a key to our success, helping to organize trips to locations as varied as the Aru Islands and Lake Habbema high in the mountains of central Papua.

The Red Birds-of-Paradise displayed only briefly, right around daybreak. I have vivid memories of making the long rope climb in the pitch dark of a moonless predawn, as stars shone brilliantly through the tree crown. I climbed into position before first light and waited for the birds—and they didn't disappoint me. On Batanta, on a rare cloudless morning, I photographed the Red Bird-of-Paradise illuminated by the first rays of the sun. —TL

At our camp, Tim downloads and reviews pictures of Wilson's Bird-of-Paradise, while Ed and guide Kris Tindige look on. Wailebet, Batanta, 17 November 2004

The Red Bird of Paradise offers a remarkable case of restricted range, being entirely confined to the small island of Waigiou, off the north-west extremity of New Guinea, where it replaces the allied species found in the other islands.

ALFRED RUSSEL WALLACE, *THE MALAY ARCHIPELAGO*, 1869

As the sun rises, a male Red Bird-of-Paradise *(Paradisaea rubra)* poses on his preferred display perch at the top of the canopy. Wailebet, Batanta, 24 November 2004

Wallace's account in *The Malay Archipelago* (1869) included this illustration of the Red Bird-of-Paradise.

Koko-o, Crater Mountain, 1 October 2005

Crossing a bridge made by locals out of small trees tied together with vines, Ed returns to camp after a long afternoon spent sitting in a blind.

Traversing the Landscape of Paradise

New Guinea is one of most difficult landscapes on Earth to traverse. Millions of years of evolution support this claim, because many closely related species, such as birds-of-paradise, live in nearly complete isolation from their nearest relatives, separated by deep ocean trenches, huge mountains, vast canyons, and massive river basins. The arduous terrain has also promoted human cultural and linguistic diversity. The New Guinea region harbors approximately 1,100 recorded languages, which makes it by far the most linguistically diverse part of the globe.

Because the island of New Guinea is divided into separate geopolitical entities (Indonesia in the west and Papua New Guinea in the east), overall statistics are difficult to obtain. But examining the information for Papua New Guinea alone, we can get a relative sense of the island's makeup. According to 2011 data, Papua New Guinea, which is about the size of California, ranks 55th (out of 250 countries tallied) in total area. Yet with just over six million people, it ranks 106th in world population, which means it's sparsely populated—again influenced by its rugged and largely inaccessible landscape. Of that small human population, only 13 percent lives in urban centers. The country ranks 136th in the world for kilometers of roadways, and among the few roads it does have, more than two-thirds are unpaved dirt tracks. This is not surprising, given the challenges of building and maintaining roads in such a rough and dynamic landscape, prone to earthquakes, mudslides, volcanic eruptions, and unstable water-soaked ground. Perhaps the most revealing fact about the island's landscape is that Papua New Guinea ranks 12th in the world for number of airports! But a closer look reveals that 96 percent of them are unpaved "grass" airstrips, which are often the only connection to the outside world for the villages that have them. New Guinea is tremendously challenging to traverse by any means: on foot, by road, and even by plane. The gates of this paradise, it seems, are truly bounded by the navigational travails of hell.

P2-WOW

Sombom, Huon Peninsula, 18 October 2011

Leaves, twigs, and forest debris swirl in the tremendous downdraft of a helicopter that has come to pick us up after two weeks in this remote area. Our gear is piled high and ready for loading at this helipad, carved out of an old unused garden plot in an otherwise uninhabited area, where roads are few and far between. *(left)*

Karawari River, Sepik region, 12 August 2005

The bright green scar of a grass airstrip stands out in marked contrast to the forest and sago swamps surrounding it. Landing in light aircraft on these remote runways is a regular part of working in New Guinea, where roads are few and far between. *(below)*

Upper Syoubri, Arfak Mountains, 3 September 2009

Navigating through a sea of dense vegetation, Ed carefully makes his way downhill through the foggy forest of the Arfak Mountains, wet in the wake of an afternoon rain shower. *(following pages)*

After years of dreaming about getting a wide-angle view of birds-of-paradise overlooking the canopy, Ed and I had finally found what might be the ideal spot for the shot.

VIEW FROM THE CANOPY

TREE CLIMBING FOR PHOTOGRAPHY

Ants swarmed over me in droves as I tried to get comfortable on the small platform I had rigged high in a tree in the swamp forest of Nimbokrang. I had climbed in the dark to a spot where I hoped the elusive Pale-billed Sicklebill would return. The day before, Ed had spotted both a male and a female feeding on a fruit-bearing vine high on the adjacent tree, and after the birds left the area, I had climbed up and rigged a small platform. Birds are creatures of habit, and a good food source often brings them back.

Stationing myself high in the tree would make all the difference, because I would be close enough to get a decent shot. The extra effort is especially important because many birds-of-paradise not only feed but also display up high in the trees. Display sites, if you can find them, are the most reliable places to photograph birds-of-paradise. If you remain on the ground, shooting photos is usually a hopeless endeavor. Ascending to the birds' level is essential, which is why tree-climbing equipment is a standard part of my kit on every bird-of-paradise trip.

I use a minimalist approach to photographing in the canopy. I don't usually erect scaffolds or construct large platforms, but prefer a small makeshift seat or platform wedged into a suitable fork in the trunk or mounted above a large branch. That way, I am less likely to disturb the birds, and I can also prepare to shoot a lot more quickly. All my gear for tree climbing and blind building fits in one large duffel. I take a bow and arrows for shooting a line over a branch high above. To fit it in the bag, I use a bow that breaks down into three pieces. I need haul lines to pull up my rope, and a harness and ascenders to climb the rope. Hunter's leafy camouflage material covers the blind so I can hide in plain sight. A few hours is usually sufficient to launch an arrow trailing a line over a high branch, pull up my climbing rope, climb, and construct a small blind.

The critical decision is choosing the right place to put it. Not every display or feeding tree has a convenient companion tree for hosting a canopy blind at the desired distance away, so lots of scouting is needed to find a suitable location. Sometimes, I must climb more than one tree to determine the one with the best view. Often, I am left with less than ideal choices, such as the incredibly ant-infested tree at the Pale-billed Sicklebill feeding site. In that case, we had been searching for a site to photograph this species for two weeks, and this was the best place we

Only a canopy platform *(opposite)* allowed such a close-up of a preening male King Bird-of-Paradise. Oransbari, Bird's Head Peninsula, 31 August 2009

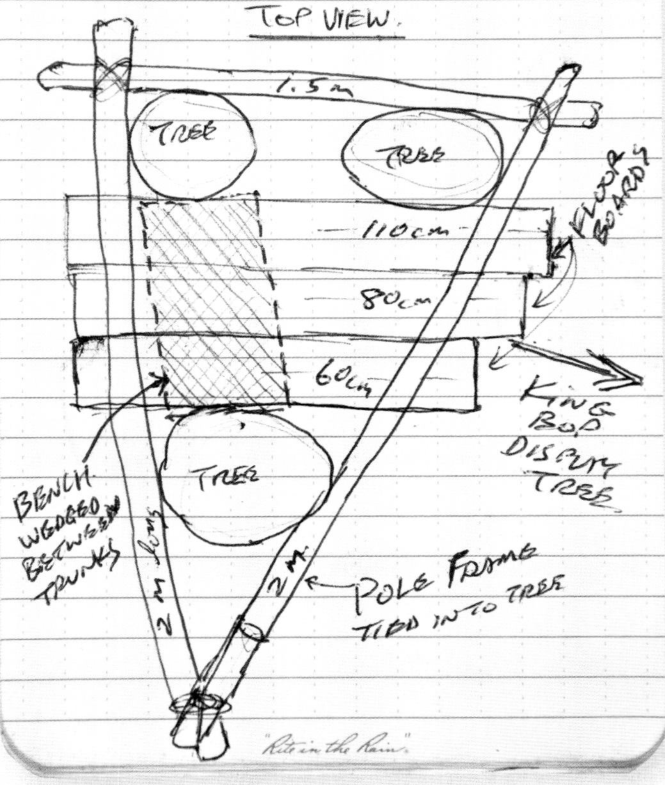

After climbing the tree, Tim sketched his platform design, including the materials to be hoisted by rope and pulley.

Tim wedges a board between branching tree trunks for a seat. Keeping oneself and everything else tied to safety ropes is essential. Oransbari, Bird's Head Peninsula, 25 August 2009

Unlike other birds-of-paradise, the King Bird-of-Paradise was not at all disturbed by his presence, so Tim was able to uncover his blind. Oransbari, Bird's Head Peninsula, 25 August 2009

could find. Spraying mosquito repellant all over the branches around me helped slow the ants down, but I was still pretty miserable. Plus, they were annoying when they crawled across my camera lens or viewfinder. But the effort did pay off. The sicklebills returned to feed, and I added another species to the collection.

THE LEAF-CAM

At 4:30 on a moonless, pitch-black morning, I had already scaled to the top of a tree in the Aru Islands. I wiped the sweat from my forehead, pulled my camera out of my backpack, and mounted it to the bracket I had attached to the tree the day before. Then I cloaked it in a wrapper of leaves and by the light of my headlamp sewed it together with rattan fiber. Next I connected the cable strung over from the neighboring tree, where my blind was located. I had to set up the camera, rappel back to the ground, and climb the other tree to my blind before the sky even started to lighten. I didn't want the Greater Birds-of-Paradise, surely sleeping somewhere nearby, to be disturbed at all.

Rappelling carefully down in the darkness after getting my camera set, I walked over to the neighboring tree where my other rope was set. With my hard case containing my laptop slung across my shoulder, I proceeded to make my second predawn climb. An immense opportunity awaited me. After years of dreaming about getting a wide-angle view of birds-of-paradise overlooking the canopy, Ed and I had finally found what might be the ideal spot for the shot, at least if the equipment all worked, and the birds weren't bothered by a camera positioned so close.

I had been trying other versions of this shot for years. The most successful previous attempt had taken place at an Emperor Bird-of-Paradise lek in the Huon Peninsula. The lek tree had no near neighbors, and the view was completely obstructed from the ground. It had been another frustrating expedition with very few good pictures, and it was literally the last day at that location when everything came together. In 2007, my cameras didn't have the ability to be controlled by a cable and laptop yet. So I had climbed the tree, rigged up a hidden camera prefocused on where I thought the birds would be displaying, and attached a remote control. Sitting on the hill below the tree, Ed and I watched with binoculars, and when we saw birds arrive and start to perform, I pushed the button on the remote and hoped I was getting a decent picture. Climbing the tree that night to retrieve the camera, I knew the entire success or failure of the expedition depended on what was stored on that memory card in the camera. Had the camera fired? Was the exposure right? The focus? The anticipation as I climbed my rope was almost unbearable. So you can imagine my elation when a quick review on the camera, while I was hanging at the top of the tree in the gathering dark, showed some very satisfying images of Emperor Birds-of-Paradise displaying upside down (see pages 44-45).

Now here in Aru in 2010, I had the next level of technology. It was amazing to see the live image from the "leaf-cam" on my laptop. Needless to say, as the sky lightened that morning and the birds came and started to display, I was incredibly excited. My jury-rigged contraption was actually working! The sun cracked the horizon. A Greater Bird-of-Paradise spread his wings as he perched overlooking the canopy. I clicked the shutter and knew I had captured a magical image (see pages 2-3). At moments like that, all the strenuous effort is forgotten. —TL

A male Greater Bird-of-Paradise with plumes held high, facing a female-plumaged bird who has just arrived, was captured with the leaf-cam, a camouflaged camera.
Wokam, Aru Islands, 22 September 2010

To use the leaf-cam, shown here with the back leaf removed, Tim placed the camera at the right height in the display tree, connected it by cable to a laptop, and operated the laptop from a blind in an adjacent tree.

Tim could observe through the lens of the leaf-cam from inside his blind in a nearby tree and record photos and videos.

The leaf-cam is in the foreground, with Tim's traditional Aru-style hunting blind in the background *(lower left)*.

Foja Mountains, 18 June 2007

One of Earth's most rugged landscapes, some mountain ranges of New Guinea are virtual islands in the sky. Their montane habitats are isolated from the next nearest mountain by a lowland habitat as inhospitable to montane species as the sea that surrounds an island.

Islands in the Sky

Mountains are the driving force behind much of New Guinea's biological diversity, including the birds-of-paradise. They dominate the landscape, looming large on nearly every horizon. New Guinea's mountains define the weather. They feed the lowland river systems. They host a wide range of habitats, from tropical forest to alpine grassland. And perhaps most critically, they divide plant and animal populations into a mosaic of subpopulations that are unable to freely interbreed, which is a critical step in the formation of new species.

Mountains span the bird-shaped island from head to tail and include peaks over 4,800 meters (16,000 feet), which are the highest in all of Australasia. They are so high that even though they lie near the equator, the tallest peaks can be clad in snow. The highest peak, Puncak Jaya (or Carstensz Pyramid), still harbors remnant glacial ice. Snow-clad mountains may not fit into your typical image of a tropical paradise, but they serve as a reminder that New Guinea is a place like no other.

By geological standards, New Guinea's mountains are young and growing at an astonishing rate. Rates of mountain building on the island are among the most extreme in the world. The uplift is estimated to be as high as 3 meters (10 feet) per 1,000 years. With mountain building like that, Mount Everest could grow from sea level to its present height in less than three million years, which is a blink of the geological eye.

In addition to this tremendous uplift, New Guinea's biological diversity also stems from the powerful lateral movement that occurs among the many fragmented pieces of continental crust that make up much of the region. These pieces are parts of the Pacific and Australian plates, which are colliding along New Guinea's north coast.

The combined effect of rapid uplift and moving plates has produced a suite of fragmented mountain ranges that forms a part of the New Guinea landmass but is separated from the central mountain ranges by vast lowland basins of unsuitable habitat. In other words, for species that favor heights, like many birds-of-paradise, these mountains are lofty islands of isolation. In terms of their influence on evolution, they are no different than coral islands set apart by the sea. The island's restless geological forces have indeed had a profound effect. Among the birds-of-paradise, 18 percent are found nowhere else in the world but within the forests of these relatively small isolated islands in the sky.

BRONZE PAROTIA ~ *Parotia berlepschi*

Bog Camp, Foja Mountains, 23 June 2007

With isolated mountains come isolated species, such as the Bronze Parotia, known for decades only from a handful of trade skins of uncertain origins. The true geographic home of this species was unknown to science until 2005. Of the five members of the genus *Parotia,* three species are found only at certain elevations in the small mountain ranges they call home. *(above)*

WAHNES'S PAROTIA ~ *Parotia wahnesi*

Sombom, Huon Peninsula, 17 October 2011

Like its relatives in the remote Foja Mountains and the mountains of the Bird's Head Peninsula of western New Guinea, the Wahnes's Parotia, a longer tailed species, is found only in a narrow band of forest at elevations of 1,000 to 2,000 meters (3,300–6,600 feet) above sea level in the mountains of the Huon Peninsula of northeastern New Guinea. *(right)*

HUON ASTRAPIA ~ *Astrapia rothschildi*

Araurang, Huon Peninsula, 5 December 2006

Fortunately, the future of birds-of-paradise like the Huon Astrapia is protected by their extremely remote and inaccessible habitats. It has not yet become economical to cut trees for timber or clear forest for agriculture at such elevations—but that may change as global resources become scarcer.

Unlike . . . other families of perching birds, the birds of paradise do not inhabit a wide range of habitats, but are predominantly confined to rainforests and a few other densely-vegetated habitats.

—CLIFFORD FRITH AND BRUCE BEEHLER, *THE BIRDS OF PARADISE*, 1998

Today you can still fly for an hour in a small plane over certain areas of New Guinea and see no sign of human beings—no roads—nothing but forest. . . . The challenge for the governments and peoples will be how to balance development with the need to safeguard a global treasure.

CONSERVATION IN NEW GUINEA

TROUBLE IN PARADISE

Birds-of-paradise are resilient. They have survived thousands of years of hunting for their plumes by native New Guineans. They have even survived periods of intense hunting for export during the peak of the plume trade for women's hats in the West (in 1913, for instance, 80,000 bird-of-paradise skins were exported from New Guinea). How have they persevered? Their system of breeding contains a side benefit. Remember that the females do not need any help from males for rearing offspring. Because hunters target only fully plumed males, the females remain unharmed to rear the next generation. Furthermore, it takes many years for males to develop their complete set of plumes, but they reach sexual maturity well before their plumes are fully developed. In the absence of fully plumed males, females will settle for a younger mate, so the reproductive rate of the population is not threatened by removing the older males.

That's the good news. The bad news is that birds-of-paradise need forest. They can live only in forests, and New Guinea's are coming under increasing pressure. Right now, only three of the 39 birds-of-paradise are considered threatened, but seven others are approaching that dubious distinction, having been classified as "near-threatened" by the International Union for Conservation of Nature (IUCN) Red List. The biggest threat to birds-of-paradise is not the traditional hunter in the highlands who still gathers some plumes for headdresses, or even the commercial hunter shooting birds to sell as decorations for homes or offices, although there is really no excuse for that. The real threat is loss of habitat. New roads are opening up huge areas to exploitation: logging, oil palm plantations, and large-scale mining operations. So far, New Guinea has a lot of forest remaining. After the Amazon and Congo basins, New Guinea harbors the largest expanse of intact rain forest in the world—by far the greatest in the Asia-Pacific region. Logging companies have long targeted Borneo and Sumatra, but with those sources dwindling, pressure is increasing to exploit New Guinea.

Today, as Ed and I have done, you can still fly for an hour in a small plane over certain areas of New Guinea and see no sign of human beings—no roads, nothing but forest. It is truly heartwarming for a conservationist to see so much untouched rain forest and just imagine the rich web of life it harbors. The challenge for the governments and peoples of New Guinea, both in Papua New Guinea and Indonesia, will be how to balance development with the need to safeguard this global treasure.

The beginning of the end? A road has been freshly cut into the pristine forest. Arfak Mountains, 28 November 2004

SUCCESS STORIES

Conservation in New Guinea faces many challenges. In the first place, this enormous island is divided between two nations. In Papua New Guinea, the government controls very little land, and traditional land ownership is fully recognized by the constitution. This has both pros and cons for conservation. On the one hand, landowners are often reluctant to allow logging companies to level the only trees they have. On the other hand, establishing national parks or nature reserves is extremely difficult.

Indonesia, with more centralized government power, has created large nature reserves and national parks in Papua with considerable foresight. Whether these areas will be successfully protected or remain "paper" parks, however, remains to be seen. Large nongovernmental organizations such as Conservation International play an important role by working at multiple levels. They provide education at the grass roots, conduct major biodiversity surveys, and work at the government level on conservation policy. These efforts require major funding and are critical to a long-term conservation strategy for the region, and we hope they can continue and expand.

What we have found most inspiring and hopeful about conservation in New Guinea are the passionate individuals who are really making a difference for the people and wildlife. One of the most exciting recent success stories has been the establishment of the Yopno-Uruwa-Som (YUS) Conservation Area, which was brought about through the efforts of Lisa Dabek of the Woodland Park Zoo and her Tree Kangaroo Conservation Project team. They focused on the endangered Huon Tree Kangaroo (*Dendrolagus matschiei*), which turned out to be a perfect species to focus people's attention. These animals are valued for food as well as for their fur, which is used in headdresses. But hunters realized that the kangaroos were becoming scarcer, and that if this trend continued, their children would be deprived of this traditional resource. In early 2009, after more than ten years, the Tree Kangaroo Conservation Project's efforts finally paid off. Thirty-five villages of the Huon Peninsula came together in a meeting with Papua New Guinea government officials and agreed to each commit a portion of their land to form a contiguous reserve that is a historic first of its kind for the country.

In Indonesian Papua as well, some local people have realized that by protecting their forests and wildlife, they can accrue more long-term benefits through ecotourism than by hunting or logging, which are short-term gains. In the Aru Islands, for instance, Wakua villagers decided to set aside a protected forest for Greater Birds-of-Paradise (see story on page 58). In the Raja Ampat islands, diving pioneer Max Ammer has set up the Raja Ampat Research and Conservation Center and is involving indigenous communities in various tourism projects—from guiding bird-watchers to providing homestays for kayakers—that have the potential to make these people's lives better without harming the environment.

In Syoubri village in the Arfak Mountains, Zeth Wonggor and his clan, with support from Indonesian conservationists Kris Tindige and Shita Prativi, set up a guesthouse that receives a steady stream of visitors coming to see the birds-of-paradise and other birds on their lands. In this village, boys no longer carry around the ubiquitous slingshot for shooting birds that is so common throughout Papua. Zeth and the village elders have convinced residents that they will be better off keeping their forest intact and their birds alive, and it seems to be working. If this kind of thinking continues to spread, I think we can be optimistic about the future of New Guinea's forests. —TL

Three endemic bird-of-paradise species, including the Emperor Bird-of-Paradise shown displaying here, are protected in the YUS Conservation Area, named for the three main rivers—Yopno, Uruwa, and Som—of the Huon Peninsula. Gatop, Huon Peninsula, 7 August 2007

A Huon Tree Kangaroo (*Dendrolagus matschiei*) mother and joey represent another species protected in the YUS Conservation Area. Wasaunan, Huon Peninsula, 4 December 2006

The YUS Conservation Area protects a range of habitats from coastal lowland forest up to montane forest such as this. Huon Peninsula, 28 November 2006

Map of the YUS Conservation Area, comprising land contributed from 35 villages. It is a valuable conservation corridor that extends from the coastal lowlands to the high montane forests of the Huon Peninsula.

Extreme Selection

When we behold a male bird elaborately displaying his graceful plumes or splendid colours before the female. . . . it is impossible to doubt that she admires the beauty of her male partner.

—CHARLES DARWIN, *THE DESCENT OF MAN*, 1871

Male birds-of-paradise receive the lion's share of attention for their varieties of bizarre behavior and extraordinary appearance. But the females truly deserve the credit for why their mates are regarded as the "most beautiful and most wonderful" of all birds. That's because sexual selection plays such a central role in their evolutionary history.

What is sexual selection, and why is it so important to the evolution of birds-of-paradise? We might best start by defining what it is not. In an 1860 letter to a colleague, Charles Darwin famously wrote, "The sight of a feather in a peacock's tail, whenever I gaze at it, makes me sick!" Of course, he didn't actually find peacocks' tail feathers to be distasteful. What bothered him was the exact opposite. It was their *extravagant beauty* that troubled him. How, or why, did these feathers evolve? To Darwin, the peacock's tail—like the plumes of a male bird-of-paradise—did nothing to improve the peacock's chances for survival. Nearly every gaudy feature of the peacock seemed more likely to reduce its ability to survive than to promote it. And that stark fact flew in the face of the idea of survival of the fittest, which was the linchpin behind natural selection, his famous idea.

Troubled, but not deterred, Darwin kept thinking and observing the natural world. A decade later, in *The Descent of Man, and Selection in Relation to Sex*, he devised a solution to the paradox posed by the peacock. He proposed the concept of sexual selection. It was an elegant explanation for what he felt was the biggest limitation to the concept of natural selection: the evolution of beauty in nature.

To explain beautiful animals like birds-of-paradise, Darwin stated that ornamental traits—the aesthetic "accessories" that offer little or no survival value—exist only because they enhance the bearer's *chances of winning a mate*. He called the process "sexual selection" and made a concerted effort to distinguish it from natural selection. In natural selection, the relative differences in mating success among individuals are determined by the environment. In sexual selection, those differences in survival are determined by members of their own species.

Of the two basic ways in which sexual selection works, the one that most fascinated Darwin is called "female choice." In this case, mating success among individual males is determined by the individual *choices* that females make about whom to mate with. With the birds-of-paradise, males are not occupied with any parental duties and are therefore free to mate with as many females as they can—and the choices of individual females tend to be highly skewed toward a mere handful of individual males that exhibit the traits considered the most attractive. As a consequence, the genetic makeup of subsequent generations is also skewed in favor of those few chosen males. Male descendants inherit the traits that were attractive to their mothers, and female descendants inherit the *preferences* for those traits. The male traits and the female preferences for those traits are therefore evolving in a coordinated manner. This coevolution of traits and preferences creates an internal feedback mechanism whereby the males with most attractive traits leave the most offspring and those offspring will have similarly attractive traits or preferences for a similarly attractive partner. Thus, if natural selection can be characterized as *survival of the fittest*, then sexual selection by female choice can be characterized as *propagation of the sexiest*. Darwin felt that beauty stemmed from the females' abilities to perceive, admire, and ultimately choose what they found aesthetically attractive.

It is important to note that, in Darwin's time, beauty in the natural world was widely considered as proof for the splendor of God's creation. Not long before Darwin rose to prominence, birds-of-paradise were celebrated as living proof of biblical truth. Even many of Darwin's contemporaries who accepted evolution by natural selection still thought only man had the capacity to appreciate beauty. Both in Darwin's time and well into the 20th century, the notion that female animals, such as peahens or birds-of-paradise, exerted preferences about which males they would mate with—thereby determining the evolutionary fate of the species—was met with widespread disbelief. Darwin's contemporaries, including Alfred Russel Wallace, and most biologists over the next 100 years attributed the males' lavish ornamental traits to adaptations by a species that ensured only members of their own species recognized them as potential mates. This theory is known as "species recognition." This pragmatic type of selection has the benefit of being entirely consistent with the original idea of natural selection.

By contrast, Darwin felt that female preferences are more or less unconstrained by the functional laws that typically dictate optimal form in evolution. Among birds, for instance, evolution usually optimizes wing shape in specific ways to, say, maximize maneuverability at the cost of speed. Natural selection hones this process according to the relative needs of species in their specific environments. Female preferences, however, are free from such functional constraints and therefore can be quite "arbitrary." Female preferences are limited only by their own sensory and cognitive abilities. Furthermore, no

GREATER BIRD-OF-PARADISE ~ *Paradisaea apoda*
Wokam, Aru Islands, 16 September 2010

A female scrutinizes one of two adult males that have synchronously positioned themselves in the head-down static pose that characterizes a peak moment of their courtship display. Although males draw most of the attention among birds-of-paradise, it is actually the females who are calling the shots during the performances. *(page 112)*

RIBBON-TAILED ASTRAPIA ~ *Astrapia mayeri*
Upper Pass, Mount Hagen, 14 December 2010

Little is known about the specifics of what female birds-of-paradise focus on when selecting a mate. Nevertheless, the ornaments and behaviors of the males themselves, such as the meter-long (3-foot) tail of this adult male, give us some clue as to what features the females of previous generations found attractive. *(pages 114-115)*

compromises need to be made for functional trade-offs, such as the trade-off in speed for a highly maneuverable wing. Ornamental male traits need not possess any virtues beyond being attractive. Female preferences can be determined simply by their "sense of beauty," as Darwin worded it.

One important point to clarify here is the meaning of the word "arbitrary." Male display traits aren't haphazard or random—they have to please the females. They are actually quite refined by the females' selective process, and their behaviors are carefully choreographed, precise, and coordinated. Just as assembling a series of sounds does not make music, so too an assortment of odd feathers and movements does not make a courtship display. Yet, just as we would recognize an infinite number of potential combinations as songs, there are an infinite number of potential displays. So what governs the differences in the form they take for a given species? According to Darwin, each species' females made arbitrary decisions according to what pleased their eye.

For the birds-of-paradise, the evolutionary impact of natural selection on the great differences in male traits may be negligible. The differences evolved merely because they were appealing to females in different ways at different times and places. The magnificently long tail of the male Ribbon-tailed Astrapia, for instance, didn't evolve via natural selection as a signal for species recognition. It probably didn't evolve as a handicap to survival signaling the genetic makeup of its bearer to females either. Rather, under the Darwinian view of sexual selection, it evolved as a mark of beauty that, once perceived as attractive, co-evolved with the female preferences for it. This co-evolution of male traits determined by arbitrary female preferences is central to understanding the tremendous diversity in ornamental features that we find, and celebrate, among male birds-of-paradise.

Because the long tail of the Ribbon-tailed Astrapia doesn't mean anything beyond appealing to females, in the future it could evolve just as easily in a vastly different trajectory for no other reason than because it's what females have come to prefer. If its females no longer favor the longest tailed males, it could become short, as we see in the Splendid Astrapia. Or if female preferences take another course, it could become black and wedge shaped, as in Stephanie's Astrapia. Or it could become broad and rounded and black, as with the Arfak and Huon Astrapias. You might say that the very existence of such a tremendous diversity of plumages, ornaments, shapes, colors, and behaviors in the bird-of-paradise supports the notion of sexual selection, at least in some cases, as an arbitrary process of female aesthetic appreciation. ■

MAGNIFICENT RIFLEBIRD ~ *Ptiloris magnificus*

Oransbari, Bird's Head Peninsula, 21 August 2009

In the same way the male riflebird shown here spends inordinate amounts of time and energy displaying, so too does the female spend inordinate time and energy observing. Selection that has led to extreme displays in males has also spurred the evolution of extreme observation and choice behaviors in females.

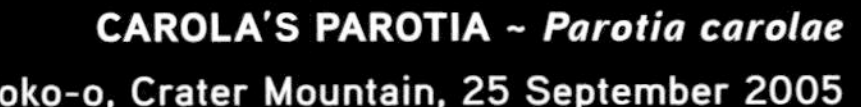

CAROLA'S PAROTIA ~ *Parotia carolae*
Koko-o, Crater Mountain, 25 September 2005

Sometimes all the females agree that one male has the "right stuff." When that one male sires many offspring and other males few or none, evolution by sexual selection is accelerated. With so many females at this court, even the neighboring males come to watch from the sidelines.

Once a male has acquired adult plumage . . . and established himself at a traditional display site he then spends much time at it. Here he displays his physical attributes with the ultimate goal of access to multiple mates.

—CLIFFORD B. FRITH AND DAWN W. FRITH, *BIRDS OF PARADISE: NATURE, ART, HISTORY*, 2010

RED BIRD-OF-PARADISE ~ *Paradisaea rubra*

Wailebet, Batanta, 23 November 2004

Getting a female to mate after watching a display from close range is only part of the challenge for any male bird-of-paradise. Even before that, a male must advertise his presence. This Red Bird-of-Paradise beckons far and wide in his effort to draw a female to the broken branch where he puts on his courtship display. *(left)*

TWELVE-WIRED BIRD-OF-PARADISE ~ *Seleucidis melanoleucus*

Nimbokrang, Jayapura area, 24 June 2010

Females of this species seem to prefer being touched by the wiry ends of the male bird's yellow plumes. Here a male is literally swiping his wires back and forth across the face of a female. *(right)*

WAHNES'S PAROTIA ~ *Parotia wahnesi*

Sombom, Huon Peninsula, 16 October 2011

Among the parotias, females not only have selected for the male's unusual plumage, complex dances, and cleared display courts on the ground but also have chosen males with at least one horizontal perch that spans the central part of the court, such as this one, which makes a convenient place from which to watch the male's displays.

A CLOSER LOOK

Finding a way to view the ballerina dance from the perspective of the female was needed to determine how the male's bizarre ornaments and behaviors had evolved.

SEEING THE FEMALE'S PERSPECTIVE

A NEW ANGLE ON COURTSHIP

From the time I first saw a male parotia performing a courtship display while a female watched from an overhanging branch, I wondered what the display would look like from the female's perspective—from above looking down. From my vantage point on the ground, the quintessential parotia display resembled a miniature tutu-clad ballerina dancing on a forest stage. The view from the ground, which was the view that I'd studied for years, is the same perspective from which the name of the display, the "ballerina dance," is derived. Yet this side view wasn't at all how females saw males when selecting a mate. Finding a way to view the ballerina dance from the perspective of the female was needed if I ever was to determine how the male's bizarre ornaments and behaviors had evolved.

Unfortunately, for years afterward my limited experience planted me firmly on the ground, observing displays from a blind. But one of the greatest advantages of the partnership between Tim and me for the Birds-of-Paradise Project is how our individual spheres of experience and expertise have helped us attain shared goals.

Upon first seeing a parotia display from the ground, Tim also wondered what the female was viewing. He was driven by the same scientific curiosity: How could we gain a better window into the evolution of this species? Luckily, Tim had special tricks up his sleeve. Nearly two decades of experience in climbing trees in the rain forest for research and photography has led him to regard challenges like this one quite differently than I do. Actually, we had *two* challenges: We had to find a way to photograph a parotia from a never-before-seen vantage point, and we also wanted to employ the tools of visual documentation to obtain new scientific information, and possibly new insights, about bird-of-paradise behavior and evolution. We agreed that the opportunity was potentially too rewarding to pass up. So for Tim, it wasn't a matter of *how* to document a parotia male from the female perspective, but *when.*

After years of photographing and documenting all of the *Parotia* species in the traditional way—from a ground blind—we decided it was time to push the frontiers of photography and scientific visualization. In 2011, on our last expedition, we attempted to see the female perspective for Wahnes's Parotia, a species endemic to the mountains of the Huon Peninsula in northern Papua New Guinea.

To do this, Tim devised a way to deploy a remotely controlled camera in a nearby tree

A female Wahnes's Parotia on a horizontal perch peers down at the male on his court below.

CAPTURING THE PEAK MOMENT ►

Each pair of photos shows the same moment of the male's display before (Time 1) and during (Time 2) the upward flash of the breast shield. Sombom, Huon Peninsula, 16 and 17 October 2011

TIME 1

TIME 2

CAMERA B VIEW

A wider view shows more of the court area, including the horizontal perch from which the females observe the male display.

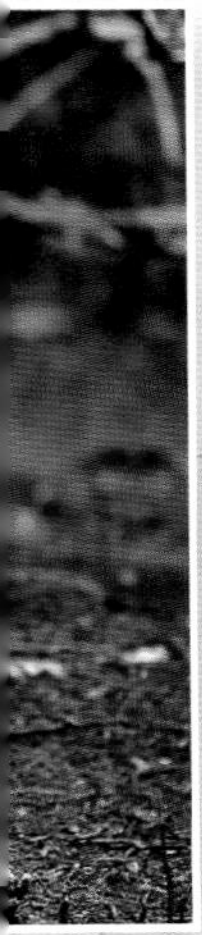

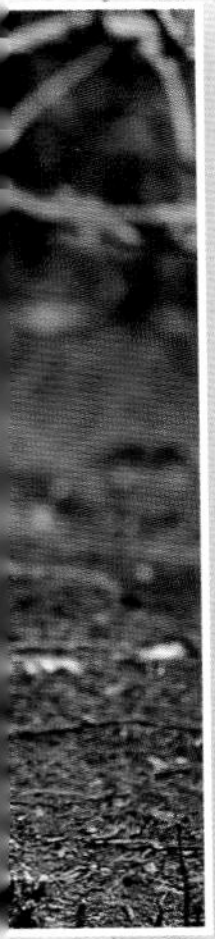

TIME 1

CAMERA C VIEW

This is one of the first images ever captured from the perspective of a female parotia. The intensity of the upward-flashing breast shield as well as the small patch of highly reflective feathers on the back of the male's head were an unanticipated surprise.

TIME 2

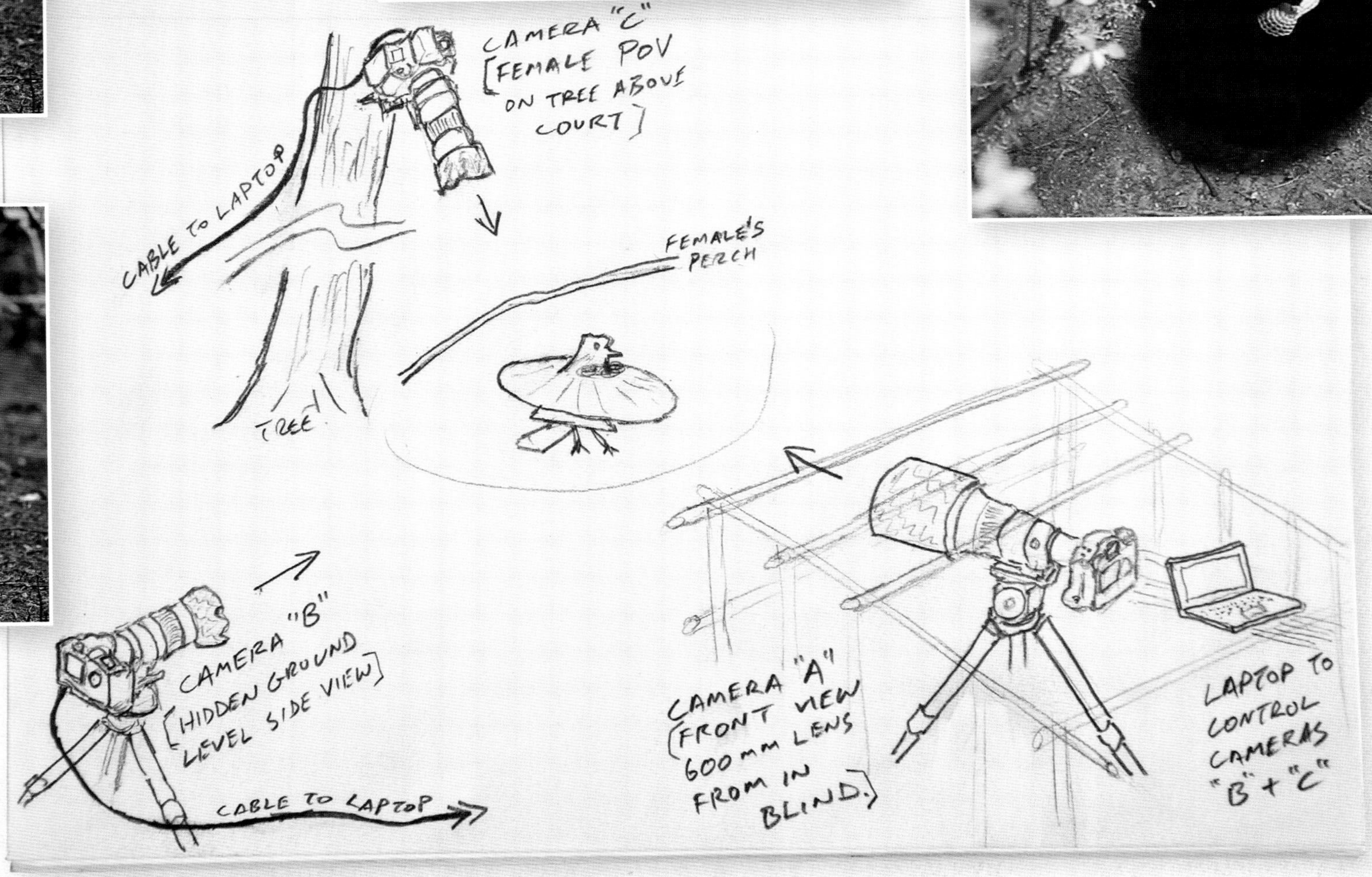

Tim's sketch shows how we set up three cameras to record the parotia display simultaneously from different perspectives at Sombom, Huon Peninsula, in October 2011.

TIME 1

TIME 2

CAMERA A VIEW

This is the traditional view of a displaying male parotia—a view that inspired some to name this particular display the "ballerina dance."

with a view of the court below. We hoped the view would correspond to the view of the female perched on the horizontal branch above the male. We found a court with a tree that had a clear view, and we built a blind on the ground big enough to accommodate both of us. Tim manned his camera with a long lens while I used a laptop to control two synchronized remote cameras: a wide-angle from the ground and one approximating the female's view. The latter Tim had to carefully conceal in a bundle of leaves about ten feet above the court. Not disturbing the birds was paramount, and we took every precaution. We employed local field assistants to make a bamboo "bush ladder" so that Tim could quickly scale the tree in the darkness of predawn to rig the camera, rather than climbing ropes, in order to minimize disturbing the birds. We unfurled more than 60 meters (200 feet) of camouflaged cables and strung them out through the forest from the camera to the blind in what we called our "bush Ethernet."

After weeks of effort, we came down to the final hours before we had to pack up our camp for good. Gloriously, the results were far better than we'd hoped. Not only did we see a displaying male as a female would see him, but we also made a surprising new discovery. Although we knew the ballerina-displaying male would look different from above, we were amazed at just *how different* it was. The name "ballerina dance" no longer seemed accurate. The male was enacting more of a "wobbling ovoid" display! We also learned how the male's specialized iridescent feather patches were used. The breast shield, which is hidden from view during most of the display, makes an upward focused flash of brilliant yellow when the male plunges his head down into his body at the start of the side-to-side head-waggle phase of the display. The biggest surprise, however, was that a relatively minor iridescent feather patch on the back of the head became prominent during the head-waggle phase. When viewed from above, it seemed to serve as a visual cue tracing the side-to-side head movement. Seeing hundreds of these displays from all the *Parotia* species, we had never understood the significance of this little ornament. This revelation was one of those much celebrated, but infrequently obtained, "Aha!" moments in wildlife photography and field research.

As fun as these discoveries were, their importance went far beyond satisfying our personal curiosity. Displays and other ornamental features of any male bird-of-paradise, not just a parotia, can be fully understood only when observed within the specific context in which they evolved. For courtship displays like the ballerina dance, the key context isn't the ecological environment. It's the "sensory environment" of the female. Displaying male parotias, and all birds-of-paradise, are essentially the physical manifestations of the cumulative individual acts of selection made by the generations of females that have come before them. So to see a male displaying the way a female would see him helps to unlock the mystery of how the species evolved. —ES

To reach the place in the tree where the "C" camera was deployed, Tim climbs the "bush ladder" our local assistants built out of bamboo and sticks tied with vine. (Photo by Edwin Scholes)
Sombom, Huon Peninsula, 17 October 2011

We hunkered down inside the "spacious" fern-walled blind built for two. In the many hours spent waiting for a bird to show, we read e-books on our smart phones. Tim reports it takes 11,564 finger swipes to read *The Count of Monte Cristo!* Sombom, Huon Peninsula, 15 October 2011

Our complicated setup involved cables, cameras, a laptop, and sound-recording gear, and keeping the equipment dry and operational was a huge challenge. Here Ed pulls up the "C" camera view on the laptop. Sombom, Huon Peninsula, 15 October 2011

Tim erected this highly camouflaged camera setup in the tree over the court to capture the perspective of a female Wahnes's Parotia. When in use, the camera was fully hidden in a mass of epiphytic ferns. Sombom, Huon Peninsula, 17 October 2011

GREATER BIRD-OF-PARADISE ~ *Paradisaea apoda*
Wokam, Aru Islands, 19 September 2010

Face to face, this female looks on as the courting male begins his final stage of display. For this pre-mating "dance," the male gets closer and closer until his chest rubs against the female and he can clap her body with his wings while pecking at her nape.

Shall We Dance?

In the world of the birds-of-paradise, the rules of attraction are created and controlled by females. The ostentatious males, with their flamboyant plumes and outlandish displays, often steal the show, but as we've seen, it's the females who call the shots. Part of the challenge for the females is controlling how and when they can be approached. They, not the males, must make the decision to mate, and so they have developed safeguards that have affected the very nature of the courtship rituals.

Male birds-of-paradise aren't picky, and more often than not, any female will do. The sole purpose of male ornaments and behaviors is, put simply, to attract as many mates as possible. The mere approach of a vaguely female-looking bird can elicit a courtship performance from a male bird-of-paradise. The process of sexual selection over millennia has finely tuned males to be little more than, to borrow a phrase from James Brown, a "sex machine."

The females have a lot more at stake in the mating game. As single-parents-to-be, the investment in selecting a mate and mating is a substantial one. Since the females won't receive any help from their partners—such as nest site preparation, food, or protection from predators and competitors—their sole concern is that their descendants reproduce. Thus female birds-of-paradise must evaluate all the males within their home range in what amounts to a comparative shopping spree during each breeding season. Males that pass the test from afar are then inspected at close range.

As the dance begins, the risks for the female become even greater. Because males are so indiscriminate, they can be overeager and try to force themselves on a female before she has definitively made her decision. That may be why the male adopts such compromising postures in so many bird-of-paradise courtship displays. From these unwieldy positions, a male cannot turn around, or turn right side up, or lower the cumbersome feathers that impede free movement, to approach a female with any degree of speed. If he tries to make a move, she has ample time to respond. Females have, over time, preferentially selected for behaviors that give them the opportunity to appreciate the male's attributes while simultaneously minimizing the risk of unwanted advances. So when a male bird-of-paradise asks, "Shall we dance?" the answer is clearly, "At a safe distance only—a very safe distance."

GOLDIE'S BIRD-OF-PARADISE ~ *Paradisaea decora*

Sebutuia, Fergusson Island, 30 September 2011

Mesmerized by the male's red plumes, this female nicely illustrates the process of female choice. Short of climbing on his back, she can't get any closer to the male, who is already touching her with his tail. Does this species' courtship have an important tactile component, as with the Twelve-wired Bird-of-Paradise? Until this photograph, there was no reason to think so.

VICTORIA'S RIFLEBIRD ~ *Ptiloris victoriae*

Wooroonooran National Park, Australia, 3 September 2008

Thwaack-thwaack . . . The females of all three riflebird species are greeted with an impressively loud wing noise as part of the male's display. How exactly the sound is produced is still a mystery, but clearly the shape and structure of the wing feathers must contribute to making the *thwaack* sound each time the male snaps his head from side to side. *(left)*

MAGNIFICENT BIRD-OF-PARADISE ~ *Cicinnurus magnificus*

Lower Syoubri, Arfak Mountains, 8 December 2004

Craning her neck to get a better look, this female is the epitome of extreme selection in action. Meanwhile, the male holds himself awkwardly upright with the specialized straw-colored feathers of his nape encircling his head like a halo. From the side, the male's front appears almost black, but from the female's point of view, it looks brilliant metallic green. *(right)*

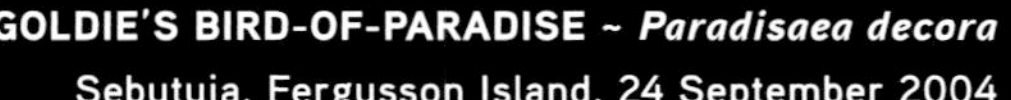

GOLDIE'S BIRD-OF-PARADISE ~ *Paradisaea decora*

Sebutuia, Fergusson Island, 24 September 2004

While courtship can take hours—perhaps even days or weeks in some cases—the act of mating in birds-of-paradise, as in most birds, is a minor affair rarely lasting more than a few seconds. Female birds-of-paradise signal their willingness to mate by gently fluttering their wings, leaning forward, and cocking their tails slightly.

Be it simplistic or complex, an individual bird's courtship and breeding strategy is directed toward one goal—the successful production of offspring.

—CLIFFORD B. FRITH AND DAWN W. FRITH, *BIRDS OF PARADISE: NATURE, ART, HISTORY*, 2010

A CLOSER LOOK

The period of prolonged adolescence before the rigors of adulthood allows males to hone their display skills through practice.

PRACTICE MAKES PERFECT

REFINING THEIR COURTSHIP MOVES

One of the most ridiculous sights in the otherwise serious business of sexual selection is watching young males practice courtship displays. What is going on here? Why do male birds-of-paradise "practice" their displays?

Because the selection process imposed by females is so exacting, males face many challenges: competition for display territories, prolonged courtship seasons, and intense scrutiny from potential mates. A male bird-of-paradise may seem to enjoy a life of leisure, but that impression discounts their considerable physical and behavioral stresses. For this reason, male birds-of-paradise have evolved several strategies to help them prepare for the rigors of adulthood: delayed plumage maturation and practice behavior.

Delayed plumage maturation refers to the fact that young males, born with the brown feathers of females, remain in "female dress" beyond their first year. In the birds-of-paradise, this stage can last up to five years and possibly more. This means that males may spend a quarter of their lives "dressed" as females before growing their elaborate male plumes. But why?

Essentially, the young males need that amount of time to maximize the chances that they will attract mates when they finally lose their female feathers. As long as they continue to resemble females, adult males don't view them as competitive threats. Among the perks of this strategy is the access they can gain to display territories in ways not possible for adult competitors. Young males have a chance to observe displays and responses of females before having to maintain and defend their own territories and display themselves.

The period of prolonged adolescence before the rigors of adulthood also allows males to hone their display skills through practice. These female-plumaged males visit the display territories of adults when the owners are away and practice the behaviors they will use when fully adult. These practice behaviors are common among most species. To the human observer, they are also quite entertaining. The young males appear inexperienced and awkward—like ungainly teenagers, which they basically are. Practice displays can also look comical because the young males behave as if they have the plumes of an adult, but they don't. You can't help but feel that you're watching someone strutting their stuff in their underwear without their knowledge. Combined with the often awkward clumsiness, the overall effect can be quite hilarious. But to the male bird-of-paradise, it's all very serious indeed. —ES

This young male Western Parotia in female plumage is using the court of the adult male *(shown below)* to practice in the master's absence. Without the feathers of an adult, the overall effect is hardly the same. But this view of the young male does reveal which feathers are involved in creating the "skirt" of an adult. Arfak Mountains, 29 and 30 November 2004

This is the same pose and same perch as above, but featuring a very different bird. The differences in appearance between a female-plumaged young male Paradise Riflebird and a fully plumed adult are striking. Australia, D'Aguilar National Park, 23 November 2010

Two young male Paradise Riflebirds in female plumage take turns playing the role of male and female while practicing the open-wing display they'll use to court a real female someday.

BLUE BIRD-OF-PARADISE ~ ***Paradisaea rudolphi***
Tigibi, Tari area, 18 September 2006

The Blue Bird-of-Paradise is not the only species to display by hanging upside down from its feet, but its distinctive inverted courtship behavior is one of the most beautifully bizarre sights in all of nature.

Bizarre Behavior

The courtship behaviors of male birds-of-paradise rank among the most bizarre of any behaviors in the animal world. Many are difficult to describe because the postures attained are so unusual and the ornamental feathers are so outlandish, but we can pick out a few common themes. Hanging upside down; turning from side to side; bouncing up and down; crouching, squatting, bowing, or leaning deeply to one side or the other; transforming their body shape by expanding feathers, or wings, into abstract geometric forms; tail swishing; head shaking; and even open-mouth "gaping" are behaviors seen again and again in different combinations.

Why such bizarre behaviors exist in the first place, however, isn't readily apparent. One possibility is that these behaviors serve the same purpose as having a model walk down the catwalk during a fashion show as opposed to simply displaying the same clothes on hangers. To fully appreciate the intricate beauty of the clothes, they are better viewed from multiple perspectives, in three dimensions, in the context for which they were created (i.e., being worn). I am speculating, but the same principle could be true for birds-of-paradise. The bizarre behavior of males might have evolved as a means of highlighting the fashionable feathers and ornaments for the onlooking females.

But it's also possible that the exact opposite is true: The ornaments might have evolved to accentuate the behaviors. This could be called the "figure skater hypothesis." Figure skaters wear ornamental costumes, but they are not meant to overshadow the movements of the performer. The best figure-skating costumes are the ones that enhance the audience's appreciation of the skater's movement without distracting attention from the athletic abilities of the performer.

Or perhaps the bizarre behaviors are a combination of both of these ideas. Call this the "carnival hypothesis." The outlandish costumes and the dance moves evolved together and play off each other to produce a cumulative effect of overall aesthetic appeal.

KING BIRD-OF-PARADISE ~ *Cicinnurus regius*

Oransbari, Bird's Head Peninsula, 31 August 2009

Although this is just a practice display, if a female were present, she would be watching from off to the right on the same horizontal branch. From her vantage point, the two emerald green circular disks at the end of the male's tail would be facing her, swaying back and forth over his head. *(left)*

RAGGIANA BIRD-OF-PARADISE ~ *Paradisaea raggiana*

Kiburu, Mendi area, 11 December 2010

Here we see a solo performance, but this species typically displays to females within a communal treetop lek. If a female were present, she would be observing the head-down posture of this male from above and behind, affording her the best view of the intensely colored portion of his flank-plumes. *(right)*

WESTERN PAROTIA ~ *Parotia sefilata*

Lower Syoubri, Arfak Mountains, 1 December 2004

With three flag-tipped wirelike feathers pushed forward from behind each cobalt blue eye, this adult male is nearing the climactic moment of his courtship display. If successful, he will end by jumping up from his performance stage to mate with a willing female on the horizontal perch above. *(preceding pages)*

STANDARDWING BIRD-OF-PARADISE ~ *Semioptera wallacii*

Labi-Labi, Halmahera, 24 July 2008

The extreme preferences of females of this species have shaped the normally tiny feathers on the top of the wing, near the wrist, into grossly elongated flaglike "standards." Coupled with the intense metallic green of the male's delta-shaped breast feathers, the overall appearance of this species' display is one of the most bizarre, if not quite the most beautiful. *(left)*

BLACK-BILLED SICKLEBILL ~ *Drepanornis albertisi*

Koko-o, Crater Mountain, 30 September 2005

This is the first-ever still image of the Black-billed Sicklebill, an elusive species, performing any sort of courtship display. While no female was seen, the male's behavior indicates that she would perch above the male, as his unusual display is clearly oriented upward. *(below)*

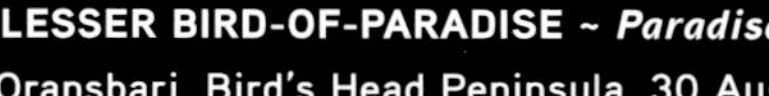

LESSER BIRD-OF-PARADISE ~ *Paradisa*

Oransbari, Bird's Head Peninsula, 30 Au

This adult male was captured in a split-se
of exaggerated plumage ruffling from
perch high in the canopy. The result is
explosion of color ar

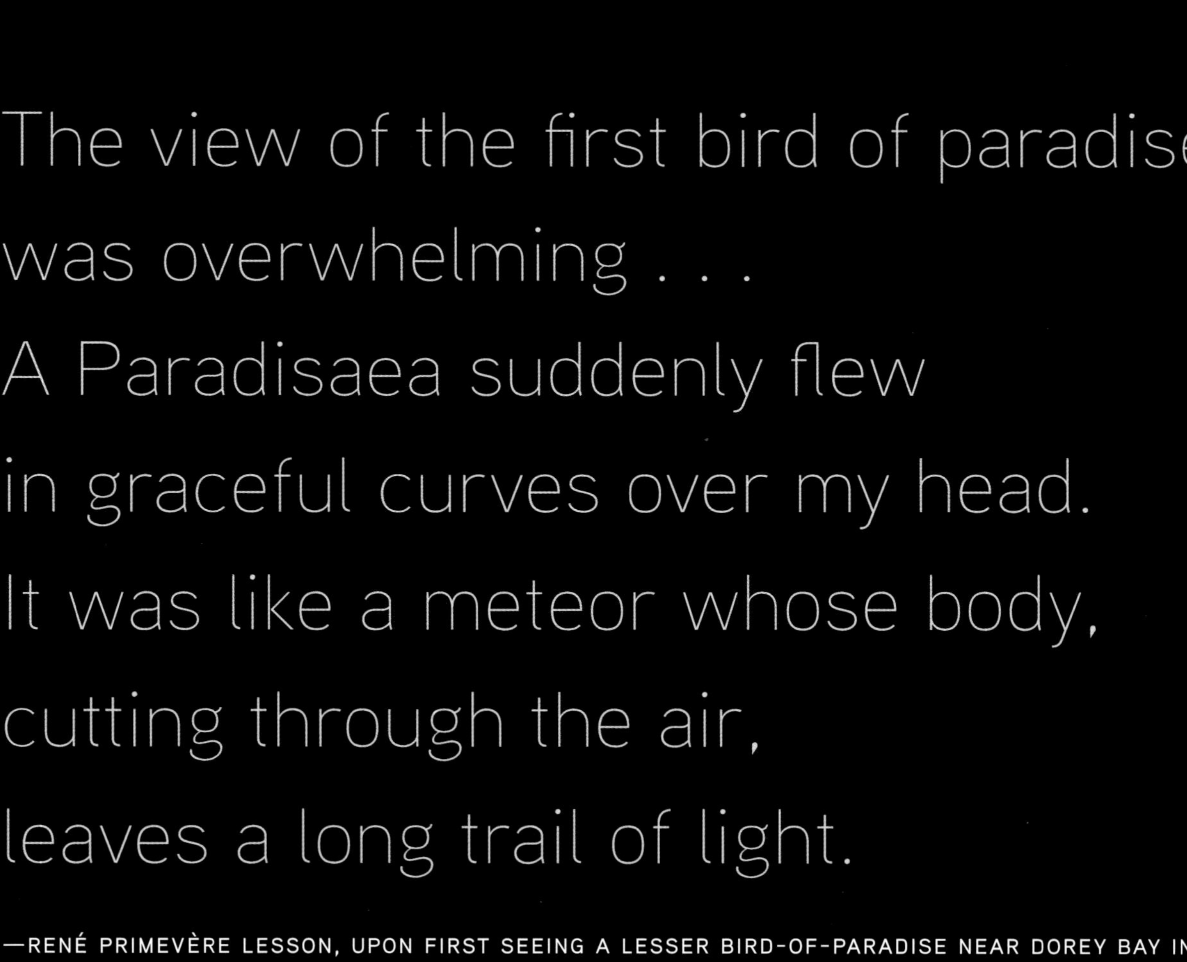

The view of the first bird of paradise
was overwhelming . . .
A Paradisaea suddenly flew
in graceful curves over my head.
It was like a meteor whose body,
cutting through the air,
leaves a long trail of light.

—RENÉ PRIMEVÈRE LESSON, UPON FIRST SEEING A LESSER BIRD-OF-PARADISE NEAR DOREY BAY IN 1824

Because of their inconspicuous nature and relative inaccessibility—most humans do not venture in these realms—most Astrapia species remain very poorly known.

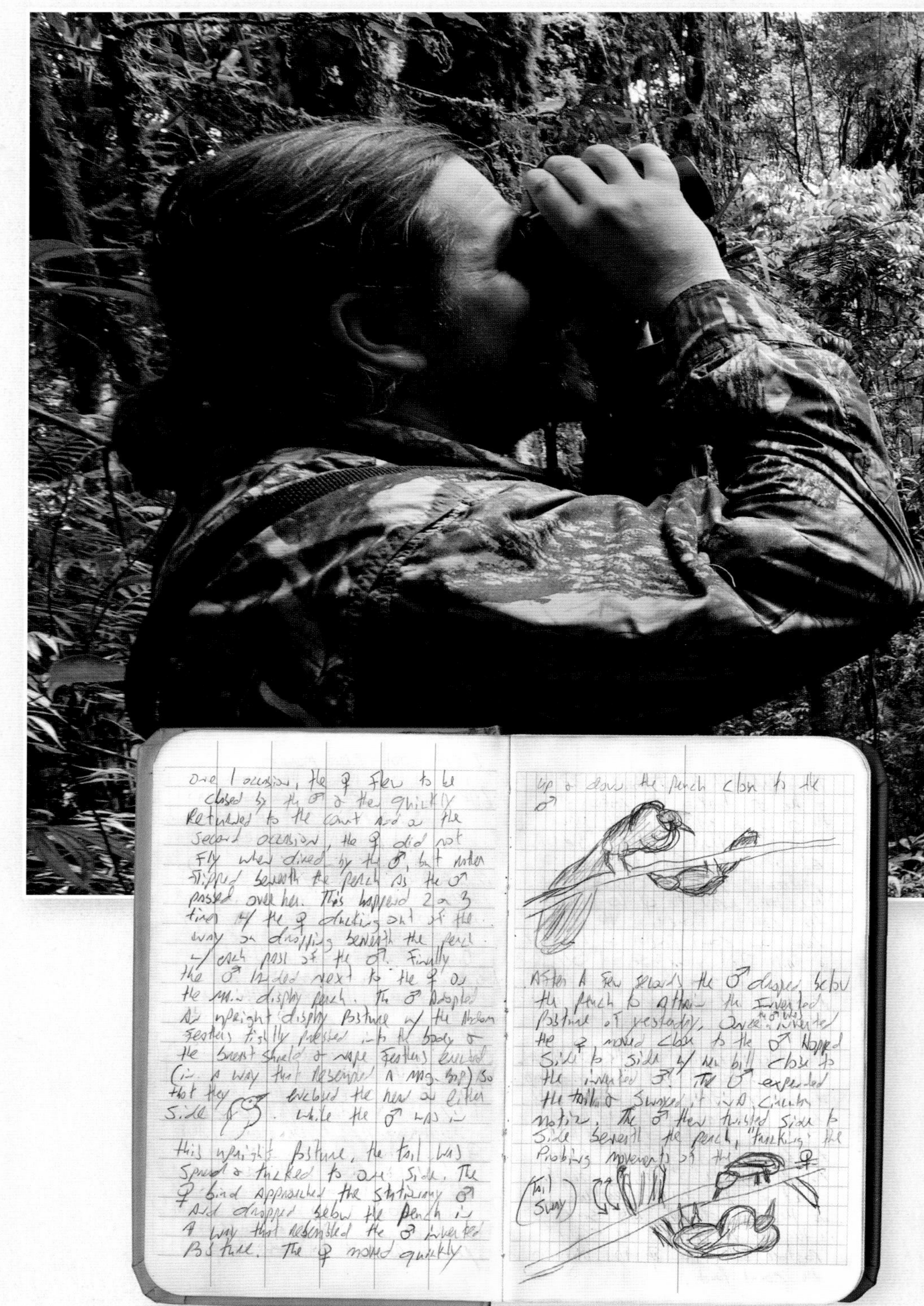

NEW DISCOVERIES: *ASTRAPIA*

REVEALING A RARE DISPLAY

Hearing the shutter on Tim's camera fire a rapid burst, I drop my binoculars and peer over my shoulder. I see him standing behind his biggest lens, which is pointed toward a moss-covered branch 15 meters (50 feet) off the ground. We are standing on a soggy cloud forest slope deep in the Arfak Mountains. I look intently trying to catch his eye and search for a sign on his face as he reviews the shots on the back of his digital camera: Did he get it? He looks up and gives a nod and a quick thumbs-up before returning his eye to the camera. He got it! Relieved, I wipe the sweat from my brow and let out a breath. With those images, the courtship display of the Arfak Astrapia became documented for science for the first time.

Nearly 15 years ago, as I was starting graduate school and dreaming of studying birds-of-paradise in the wild, I read a new monograph on them by Cliff Frith and Bruce Beehler. That book was invaluable to a beginning Ph.D. candidate interested in the topic, and I suppose I've read it cover to cover a dozen times by now. But what stands out most in my mind from that initial reading was a blank I found on page 252 in the species account for the Arfak Astrapia. In the subsection titled "Courtship Behaviour," which was usually the longest part of each species' description, was just a single word that loomed in profound solitude on the page: "Unknown." The lacuna was shocking even then, though many species of birds-of-paradise were poorly known. That one word ignited a flame of curiosity that burns to this day. One of the most exciting aspects of embarking on a career studying birds-of-paradise in the wild was that biological mysteries like the courtship dances of the Arfak Astrapia still needed to be explored.

Tim and I both have backgrounds in science, and we have worked as science professionals. From the start, we aimed not only to promote a science-based approach through our project but also to actually conduct research and use the resulting photographs and audio and video recordings for the advancement of scientific knowledge. Because our survey had the goal of being comprehensive, cataloging all 39 species of the birds-of-paradise, one particular area where we knew we might make a significant scientific contribution was within the genus *Astrapia*.

The astrapias are the mysterious long-tailed birds-of-paradise found in the higher reaches of New Guinea's mountain ranges. Their homes are cloud forests in every sense of the word—sun is infrequent, mist and fog are the norm, and

Finding a display site for any bird-of-paradise requires a lot of time spent in the appropriate habitat, observing and listening, along with lots of patience and a good dose of luck.

The display began w/ the ♂ sitting upright on the perch and flicking his wings & rapidly spreading the outer tail feathers.

After flicking wings & tail for about a min. the ♂ would lower his head below the plane of the perch & raise his tail up over his head w/ tail again spread

Following the head down tail-up posture the ♂ resumed the upright wing & tail flicking

After a few secs of being upright the ♂ suddenly adopted an inverted posture by dropping beneath the perch tail first, but also to the side

In this posture, the ♂ fully erected the breast shield & flank plumes as well as the feathers around the head so that the upper body engulfed the head & neck. At the same time, the body was elongated & lower portion was flattened. Once inverted, the body appeared to be held mostly parallel w/ the perch except for the tail which was cocked up @ a 75° angle so that it was elevated above the plane of the perch. The tail was fully fanned.

While inverted w/ tail fanned & breast plumes erected the ♂ would slightly twist side to side beneath the perch while swaying the tail in slow deliberate motions back & forth while expanding & contracting the tail feathers

This display was performed 2 or 3 times each during the 3 bouts of display. Each inverted display lasting ± 10 sec w/ about 4-5 "sways" of fanned tail.

The pages from Ed's field notebook in 2001 show when he first encountered a displaying Huon Astrapia, including his observations, illustrated, of the first-ever documented displays to a female.

An adult male Huon Astrapia slides, tail first, below the perch to turn upside down. The courtship display of this species was poorly known and not photographed before this. Araurang, Huon Peninsula, 5 December 2006

the riot of flora is thick with moss and very wet. Because of their inconspicuous nature and relative inaccessibility—most humans do not venture in these realms—most *Astrapia* species remain very poorly known. Their courtship behaviors are a special case in point. Since our documentation efforts aimed at revealing the details of bird-of-paradise courtship displays, we were well situated to make new discoveries.

One example of breaking new ground occurred with a species found only in the isolated mountain ranges of the Huon Peninsula of northeastern New Guinea. Prior to the official start of the Bird-of-Paradise Project, I'd had the chance to see the courtship display of the Huon Astrapia when doing research on Wahnes's Parotia, which is also found only in the mountains of the Huon. Over the course of a week, I was able to see numerous complete displays to females, including several that ended with matings. This was the first time anyone had seen the full display of that species in the wild, and while my scientific observations were solid, the quality of the documentation left much to be desired. So, when Tim and I began working together, we returned to photograph and document that species. In doing so, Tim was able to get excellent photographs and high-definition video of the very same behaviors that I had seen years before. Through the combination of both of those efforts, a clearer picture of the courtship display repertoire of the Huon Astrapia has emerged, and the subject is being described in a scientific paper. It's just one outgrowth of the scientific underpinnings of our project.

A second, even more exciting example targeted the species found only in the mountains of the Bird's Head Peninsula of western New Guinea: the Arfak Astrapia. The courtship behavior of this species was, until the first clicks of Tim's shutter in 2009, completely unknown and undocumented. Based on some intriguing similarities in plumage between these two species, I had once speculated that the displays of the Arfak Astrapia would be somewhat similar to those of the Huon Astrapia. At that time, the Huon species was the only astrapia known to display by hanging upside down. During courtship, the male rotates tail first to hang upside down with his body held horizontally beneath a branch and fans his flaglike tail to the female perched above. While tracking the elusive Arfak Astrapia in the Arfak Mountains, we were surprised and elated when we came upon a not-quite-fully adult, but partially plumed, male perform a series of practice displays in which he adopted an upside-down display posture nearly identical to that of the Huon Astrapia! Although the photos and the video we took aren't the most perfect and beautiful of the project, they nevertheless represent one of the most significant scientific discoveries, which in a lot of ways is even more important. —ES

The courtship behaviors of the Arfak Astrapia were unknown until this day when, with the help of local expert Zeth Wonggor, Tim captured the images shown here. Upper Syoubri, Arfak Mountains, 3 September 2009

This photographic sequence represents the first documented courtship displays of the Arfak Astrapia, a major scientific discovery and a significant contribution to the natural history of this species. Upper Syoubri, Arfak Mountains, 3 September 2009

Deep in the Arfak Mountains, Tim swivels his big lens into place while tracking a subadult male Arfak Astrapia *(shown above and at left)*, poised to capture the first-ever images of this species' courtship display. (Video frame by Eric Liner) Upper Syoubri, Arfak Mountains, 3 September 2009

Jacques Barraband's interpretation of the display pose of the Arfak Astrapia was originally published in François Le Vaillant's *Histoire naturelle des oiseaux de paradis et des rolliers* (1801–06).

Absurd Exaggerations

[B]irds of paradise raise difficult questions, questions that penetrate to the very foundation of our biological theories. How can natural selection favor, one might almost say permit, the evolution of such conspicuously bizarre plumes and displays? How can it permit such "absurd exaggerations" as one is almost tempted to call them? How can it happen that apparently closely related species and genera differ so drastically in their habits and colorations?

—ERNST MAYR, FOREWORD TO E. THOMAS GILLIARD, *BIRDS OF PARADISE AND BOWER BIRDS*, 1969

As the quote that opens this chapter shows, as recently as 1969 one of the most prominent evolutionary biologists of the 20th century still had not recognized the significance of Darwin's ideas about sexual selection. Ernst Mayr's concerns sound quite similar to Darwin's from more than 100 years earlier (see quote on page 155). Mayr indicates that the "absurd exaggerations" of birds-of-paradise present "difficult questions" that "penetrate to the very foundation" of natural selection. Even at this point in his career, Mayr could not conceive of the notion that female mating preferences could shape the evolution of a species. Nor was he alone at that time. To explain the "conspicuously bizarre plumes and displays" of birds-of-paradise, most biologists preferred Wallace's natural selection logic to Darwin's sexual selection. Another decade would pass before sexual selection would emerge as a major force in the evolution of animal diversity, but today it has become a preeminent field of study.

Mayr's comment also questions how so many closely related species can "differ so drastically in their habits and colorations," pulling our focus back to another point made by Darwin: the evolution of *biological diversity*—Darwin's "endless forms." The reason this question is important is because, when viewed as a cohesive evolutionary group, the birds-of-paradise exhibit a nearly unbelievable degree of *disparity* in appearance and behaviors. It is the substantial differences among species, more than the extreme elaborations within any one, that make the birds-of-paradise so absolutely astonishing. To think of it another way, consider what it would be like if the birds-of-paradise were a group of 39 exceedingly beautiful and fabulously bizarre, but *very similar-looking* species. They would no doubt be recognized for their beauty, but their intrigue would be much less than it is with so many exceptionally different species. In other words, that the birds-of-paradise have such "absurd exaggerations" is interesting, but that they have so many drastically different ones among such closely related species is even more interesting.

So how can we begin to characterize the diversity among the birds-of-paradise? In general terms, two broad patterns of ornamental evolution have come about through sexual selection. Although these patterns are not hard and fast—there are clear exceptions and "in-betweens"—they nevertheless help make sense of bird-of-paradise diversity.

The first pattern defines a group that includes the classic birds-of-paradise of historical fame—the plumed species for which color, and the way it is lavishly presented, is the most prominent feature of the male's courtship. These are the "color-shakers." In this group, we find the *Paradisaea* species of legend, such the Greater Bird-of-Paradise, that sparked the imagination of traditional New Guinean and European cultures alike. The colorful plumes of these species were so unprecedented that they essentially blinded people to the normal bird parts—the wings, tails, and legs—hidden among them. Instead, they spurred legends of ethereal creatures from the heavens (see page 36). The vibrant feathers of these species are so attractive as ornaments that they have been worn on the heads of people from both traditional and high-fashion cultures. The color-shaker species are quintessential examples of sexual selection crafting feather shape and structure in ways that greatly emphasize color.

The reason we have used the term "color-shakers," however, is the role that the *shaking* plays. The males of these species have an array of courtship behaviors that have evolved in concert with the colorful feathers in order to highlight them (or vice versa!). The displays of the color-shaking species include everything from synchronized group "explosions" of color from the canopy, as by the many *Paradisaea* species, to the solitary twists and twirls used in the "pole-dances" of the male Twelve-wired Bird-of-Paradise. The color-shaker group also includes the vivid riot of color flashed, fanned, and swayed for females during the displays of the Magnificent, Wilson's, and King Birds-of-Paradise. The common theme that all these species share is that color is primarily what is being flaunted during courtship. That means it is also the primary feature that females of these species favor.

The second group of birds-of-paradise does not emphasize color so much as *shape* and *contrast*. In these species, we see a *transformation in shape* from a birdlike form into a configuration that is virtually unimaginable. We call the species that have evolved these types of absurd exaggerations the "*shape*-shifters." The parotias, riflebirds, the Superb Bird-of-Paradise, and the *Epimachus* sicklebills are all premiere examples of shape-shifting species. These species have evolved a complex interaction of feather structure and coordinated behavior that alters their shape. This is not to say that the shape-shifters don't use color, but rather that their colors are mostly limited to highlighting specific points, or moments in time, within the formation of their bizarre shapes. In contrast to the color-shakers, which are bright yellow, brown,

VICTORIA'S RIFLEBIRD ~ *Ptiloris victoriae*
Wooroonooran National Park, Australia, 6 September 2008

For many male birds-of-paradise, sexual selection has resulted in absurd exaggerations that are substantial modifications from the norm. This male riflebird, for example, shows how female mating preferences have fundamentally altered the aerodynamically optimal wing shape in favor of blunt-tipped feathers that make an attractive ovoid display. *(page 154)*

KING OF SAXONY BIRD-OF-PARADISE ~ *Pteridophora alberti*
Lower Pass, Mount Hagen, 26 July 2005

A feathered wonder of the world, the male King of Saxony Bird-of-Paradise sports two antenna-like plumes on his head, unlike any other feather known to exist. They appear so absurd and unnatural, in fact, that some early collectors assumed they were fabrications manufactured to command a high price in the plume trade at the turn of the last century. *(pages 156-157)*

orange, and red, the shape-shifter males are mostly black—usually jet black—and the colors they employ are strikingly bright and iridescent blues and greens.

Without question, the shape-shifting species aren't as widely lauded as the color-shakers. They aren't depicted on flags or painted on airplanes or printed on currency. Even the publishers of this book resisted putting one of these rarely photographed species on the cover because their darkness often overshadows the astonishing spectacles they hide within. But as you delve more deeply into the rarely seen world of courtship in these species, the wonderment of the shape-shifters takes hold. The bizarre allure of the shape-shifters is subtle and enigmatic compared with the superficial beauty of the color-shakers. The color-shakers are spectacular to behold, but they appear spectacular all the time, not only during display (hence their use on hats and headdresses). But the bizarre beauty of the shape-shifters is not revealed unless you are lucky enough to see the magic that happens in the moments of transformation—the moments when precise coordination of body parts, uniquely shaped feathers, and the diffuse light of their forest understory stages all combine to create some of the most astonishing sights in all of the natural world.

The female shape-shifters have come to prefer extreme modifications in body shape over evolution of extreme colorations. As a consequence, the unusual formations presented by the shape-shifting males are directional and meant to be viewed only from the perspective of an observing female. Unlike the nearly always visible beauty of the color-shakers, the incredible courtship appearances of the shape-shifters can't be fully appreciated, or even understood, unless viewed from the precise perspective of the females—that is, from the context in which they evolved (see page 126). These secrets of the shape-shifters are, without a doubt, some of the most fascinating courtship wonders among all the birds-of-paradise.

The final section of this book is a celebration of the diversity of outcomes that sexual selection has produced during the course of bird-of-paradise evolutionary history. The many "absurd exaggerations" that make the birds-of-paradise so unusual are at the core of what makes them so interesting in the first place. In this final part of the book, we'll take a closer look at the extremes of ornamental elaboration and the array of "endless forms" that have evolved within this spectacular group of 39 closely related species. ■

RIBBON-TAILED ASTRAPIA ~ *Astrapia mayeri*

Upper Pass, Mount Hagen, 19 August 2005

With a tail so long that this male could wrap himself in it several times over, the name "Ribbon-tailed" is hardly hyperbole. Rather astonishingly, the two central tail feathers are lost and regrown annually, as is every other feather on the bird. The annual molt is timed to ensure that his feathers look their best for courtship display season.

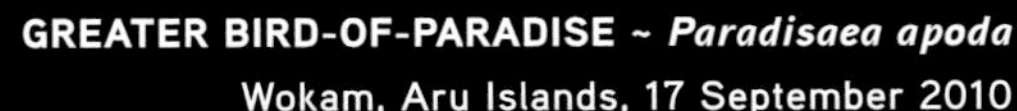

GREATER BIRD-OF-PARADISE ~ *Paradisaea apoda*
Wokam, Aru Islands, 17 September 2010

The warm light of sunrise appears to glow from within the yellow flank-plumes of these adult males who have gathered, as they do nearly every morning, to display from the branches of their canopy lek. It is these radiant plumes that led to the hunting of tens of thousands of these birds during the height of the plume trade.

The spectacular and ornate plumages of these birds in fact represent the extreme expression of the process of sexual selection.

—CLIFFORD FRITH AND BRUCE BEEHLER, *THE BIRDS OF PARADISE*, 1998

MAGNIFICENT BIRD-OF-PARADISE ~ *Cicinnurus magnificus*

Lower Syoubri, Arfak Mountains, 10 December 2004

Often difficult to see, the brilliant metallic green breast feathers of this adult male expand and pulse as he tracks the movements of a female along the periphery of his terrestrial display court. Even the legs and feet of this species (and its close relatives) have evolved into ornaments. *(left)*

CAROLA'S PAROTIA ~ *Parotia carolae*

Koko-o, Crater Mountain, 23 September 2005

Like an odd-looking whiskered man wearing a tutu, this adult male shuffles forward and backward along his cleared ground court, jerkily shaking his head from side to side. Above, out of view, a female watches the performance from a horizontal perch. This species has the most complex courtship display known among all the birds-of-paradise. *(right)*

TWELVE-WIRED BIRD-OF-PARADISE ~ *Seleucidis melanoleucus*
Nimbokrang, Jayapura area, 16 June 2010

Extraordinarily bizarre, the 12 "wires" that form the twisted mass at the rear end of this male are not a part of his tail, as commonly thought. In truth, they are the elongated central shaft, or rachis, of the bright yellow flank-plumes.

Extraordinary Ornamentation

The diversity of male ornamentation among the birds-of-paradise is extraordinary. Some have ribbonlike tails three times as long as the body. Others have wiry feathers that conspicuously protrude from the head, body, and tail. A few have plasticlike feathers so unlike the usual construction that they seem to be man-made. Many have some version of a feather-made "fan" that radiates from the chest to encircle the upper body. Others have greatly elongated plumes that transform a bird into a flowery explosion of color. At least one has a feather "cape" that both exaggerates and conceals shape. A number have incredibly colored skin, including the insides of the mouth! Nearly all have feather patches that shimmer with brilliant metallic colors. In three species in particular, even the wings have been dramatically rounded for visual effect rather than being shaped in an optimal way for flight. In male birds-of-paradise, virtually no body part has been spared the evolutionary transformation from ordinary functionality to ornamental object of desire. And what extraordinary ornaments they are.

No other group of closely related bird species has such a phenomenal array of sexual ornamentation. The peacock and its relatives are remarkable, to be sure, as are many hummingbirds, ducks, and a few other avian groups in which sexual selection has had a strong influence over the course of evolution. Yet none show the spectacular diversity or extremes of ornamental form that is found among the birds-of-paradise.

The evolution of all the shapes, colors, and textures of the feathers and other body parts, however, cannot properly be considered outside the context of the many bizarre courtship behaviors that birds-of-paradise employ. An ornament, by definition, is used to beautify appearance, but otherwise has no practical purpose. The courtship display behaviors, too, are examples of extraordinary behavioral ornamentation that serves no purpose beyond being attractive. The diversity of behavioral ornamentation that birds-of-paradise exhibit has few peers among all birds. Some display behaviors are relatively simple and others are composed of many intricate parts, but all are showy and revealing.

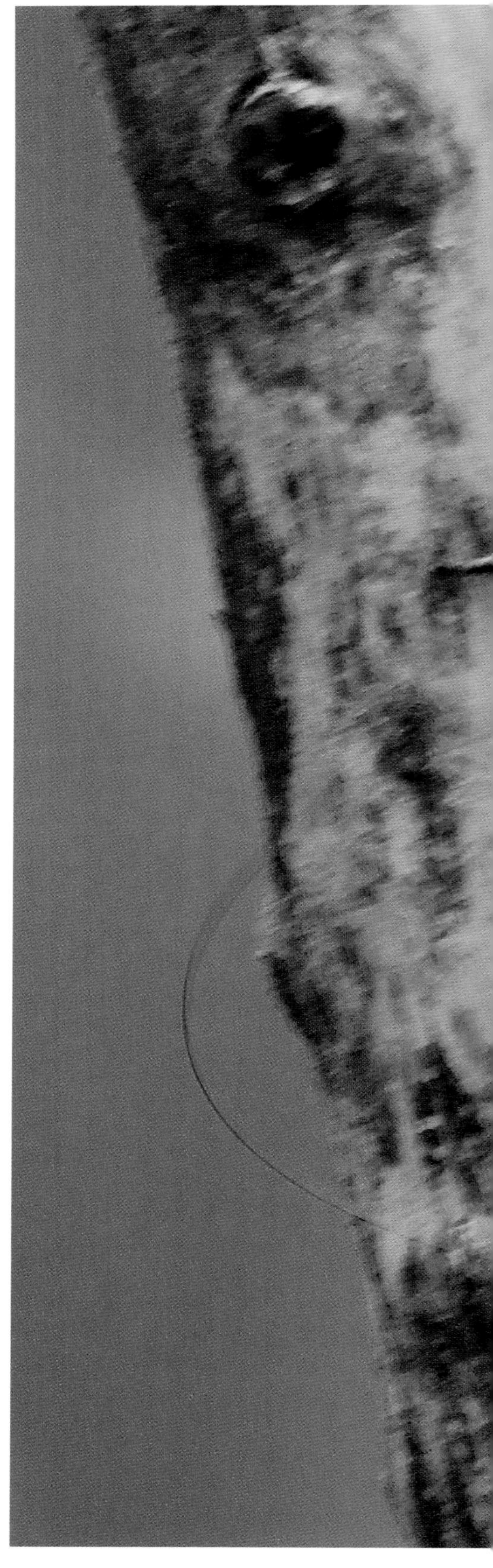

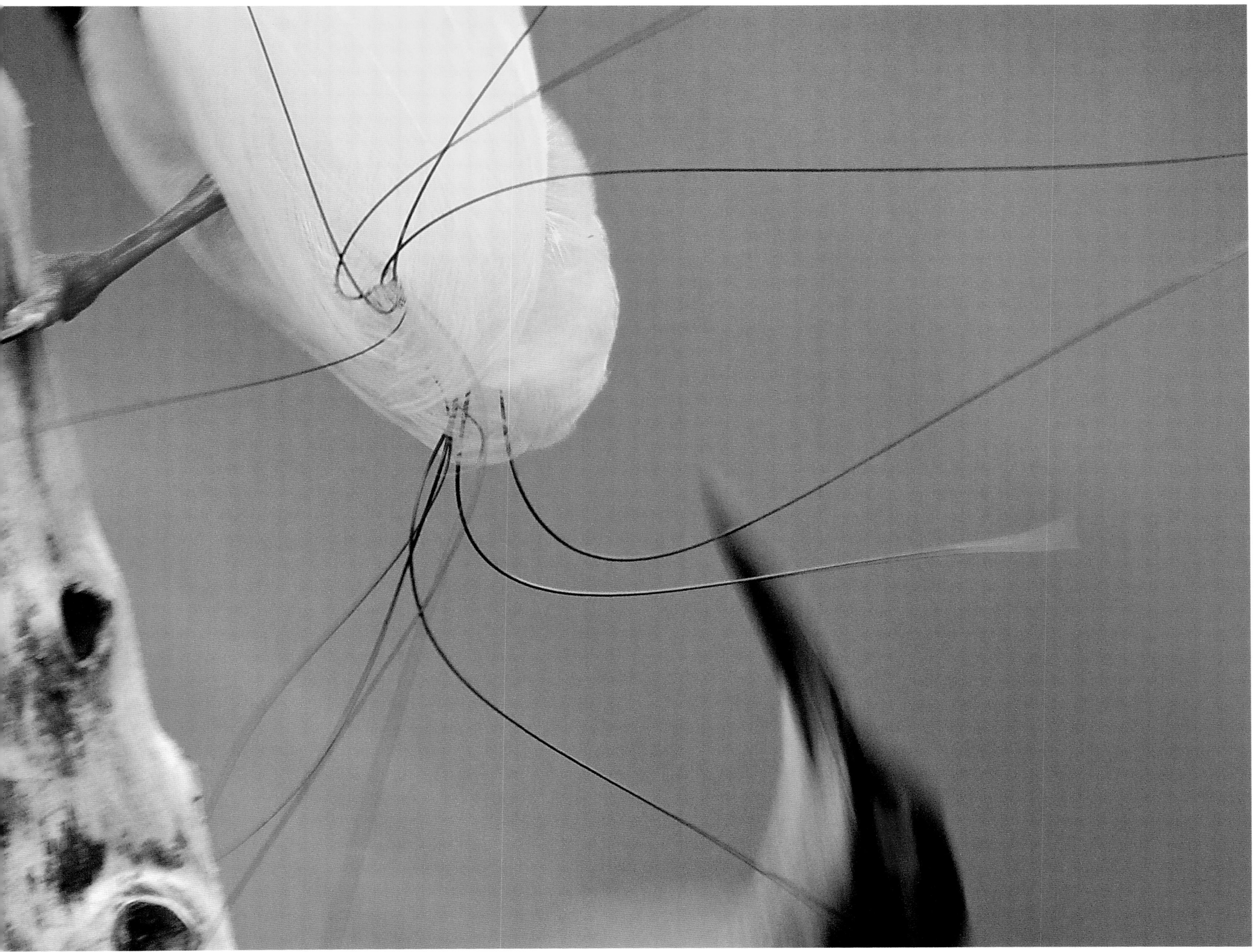

WILSON'S BIRD-OF-PARADISE ~ *Cicinnurus respublica*

Waiwo, Waigeo, 2 October 2010

The top surface of this adult male shows the same primary color palette of red, yellow, and blue that artists have used for centuries to create every other color. Add the emerald green of the breast, the deep blue of the legs, the yellow-green mouth, the purple underparts, and the metallic violet-blue of the curled tail, and this bird becomes a true cornucopia of color. *(left)*

SPLENDID ASTRAPIA ~ *Astrapia splendidissima*

Lake Habbema, Snow Mountains, 21 June 2010

Shimmering like a feathered disco ball, the plumage of this adult male is as reflective as that of any bird. Because iridescence works subtly—just the right alignment is needed between the light source and the eye of the viewer to achieve maximum effect—few people have seen the astonishing colors of this species. From most human vantage points, these birds look black. *(right)*

SUPERB BIRD-OF-PARADISE ~ *Lophorina superba*

Tigibi, Tari area, 30 October 2011

The contrast of highly iridescent feathers against a black background is a major feature of most of the shape-shifting birds-of-paradise. Even against a bright sky, the brilliant metallic blue breast shield of this adult male stands out against the deep black plumage of the body. *(following pages)*

The feathers that really stretch the imagination serve as the centerpieces of the "costumes" that males display to females during courtship.

FEATHERS OF SEDUCTION

DIVERSITY IN SHAPE, SIZE, AND COLOR

Because feathers are so familiar to us, we rarely pause to marvel at how ingeniously they are constructed. Feathers are lightweight but durable. They manage to waterproof and insulate while being as breathable as Egyptian cotton. They are also wonderfully multifunctional. Feathers help make an airfoil for creating lift on the wings, and they serve as rudders for "steering" on the tail. But perhaps the most wonderful feature of feathers is their extraordinary diversity of colors, color patterns, and their use as a means of concealment, on the one hand, or extravagant ornamentation, on the other. Equally as remarkable, each feather is lost and regrown annually through the process of molting. Even the one-meter-long (three-foot) tail feathers of the Ribbon-tailed Astrapia are dropped and regrown annually. It's not an exaggeration to say that the feather is one of the most versatile biological structures ever to have evolved.

In the birds-of-paradise, feathers became so sensational because of their function as ornaments of seduction. That isn't to say bird-of-paradise feathers don't serve all the other functions. But the feathers that really stretch the imagination serve as the centerpieces of the "costumes" that males display to females during courtship. These are the feathers that have made these birds so appealing to people ever since they first laid eyes on them. These feathers are used as ornaments among native New Guineans. They drew the hunters who wanted to meet the demands of high fashion in 19th-century Europe and America. In other words, the feathers themselves have evolved in ways that make these birds so striking.

The most famous feathers are the exceptional flank plumes of the ***Paradisaea*** species. In a typical bird, the flank feathers are just the normal feathers that cover the body. In the female Greater Bird-of-Paradise, for example, the flank feathers are only two-and-a-half-centimeter-long (one-inch), oval-shaped body feathers. In the males, however, these same feathers can be over 30 centimeters (1 foot) long. Instead of having a more typical feather shape, they are filamentous and wispy, or "plumaceous." What does that mean? The plumaceous structure comes from different microformations of the same basic parts that make up a typical feather. Ordinarily, the central shaft has a series of branches coming off in a row on each side, and each of these branches itself contains a series of many more, even smaller branches. All of these branches and sub-branches are tightly packed together and

GREATER BIRD-OF-PARADISE

VARIETY IN FEATHER STRUCTURE

From widely spaced feather barbs that create a fluffy effect *(above)* to flags, disks, wires, and ribbons, the variety of feather structure in birds-of-paradise is simply unparalleled.

STANDARDWING BIRD-OF-PARADISE

KING BIRD-OF-PARADISE

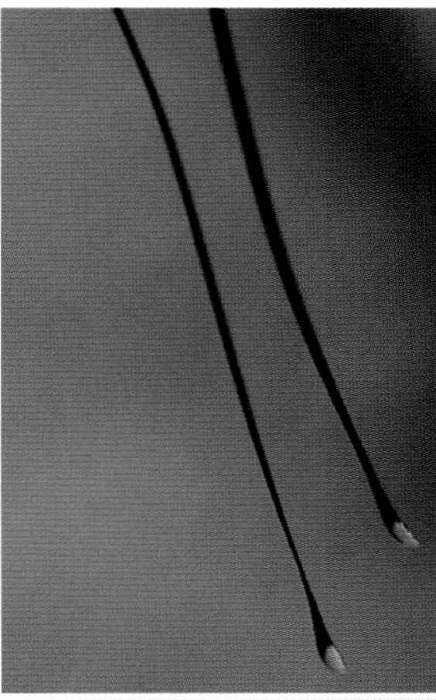

BLUE BIRD-OF-PARADISE

RED BIRD-OF-PARADISE

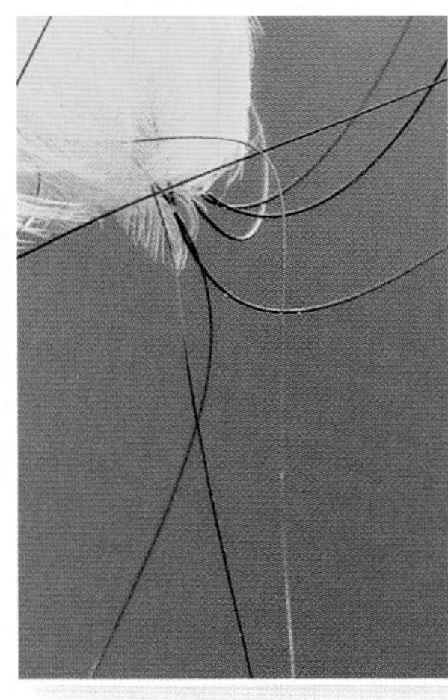

TWELVE-WIRED BIRD-OF-PARADISE

WILSON'S BIRD-OF-PARADISE

WAHNES'S PAROTIA, WITH SIX HEAD-WIRES WITH PADDLE ENDS

KING OF SAXONY BIRD-OF-PARADISE, WITH TWO HEAD WIRES WITH PLASTICLIKE TABS

KING BIRD-OF-PARADISE, WITH BREAST FAN AND VELVETY HEAD FEATHERS

actually hook onto one another like Velcro, creating the "fabric" of a typical feather. But in the plumes of the male Greater Bird-of-Paradise and its relatives, the branches and sub-branches are stiffer and more widely spaced and they don't hook together to form a smooth surface. Instead, they explode out, creating a unique firework-like pattern of fluffy color.

Besides the well-known *Paradisaea* plumes, the birds-of-paradise have other feathers with shapes and textures unlike anything seen among other birds. One repeated theme is "wires." The five *Parotia* species have wires, three per side, that stick out from the head behind each eye. The Twelve-wired Bird-of-Paradise gets its name from the flank-borne wires that form a twisted cloud at the male's rear. The King Bird-of-Paradise, along with several other species, has wires that are long extensions of the tail. All of these "wires" come from feathers in which there are no branches off the central feather shaft at all. Interestingly, the branches aren't removed by preening, pruning, or wear; they simply are never grown. In other words, these feathers have *evolved* into wires through the process of sexual selection.

One species, the King of Saxony Bird-of-Paradise, has two bizarre feathers that emerge from behind the eyes. The "fabric" of these long baby-blue-and-white, banner-shaped feathers appears and feels unnatural, like plastic or some other man-made material. These one-of-a-kind feathers have flattened and fused their branches and sub-branches, fundamentally defying what it means to be a feather.

Beyond the astonishing array of shapes and textures of bird-of-paradise feathers are their extraordinary colors. Birds-of-paradise produce their feather colors by employing the same two basic mechanisms used by all other birds. Most commonly, coloration is created by colored pigment molecules deposited inside the feathers as they grow. The many bright yellows, oranges, and reds are made by carotenoid pigments extracted from the bird's diet. The other major way in which birds produce color is the selective reflection of the different colors, or wavelengths, of light interacting with the microstructures of the feather itself. Colors produced in this way are called "structural colors." This mechanism of color is similar to the way a rainbow is colorful despite the fact that the raindrops creating it are colorless themselves. The Blue Bird-of-Paradise derives its brilliant blue color from such a mechanism (as does the more familiar Blue Jay). So, too, the brilliantly metallic iridescent breast shields of the parotias and riflebirds, as well as the luxurious flashing of the astrapias, are the product of this type of structural color.

Whether in shape, texture, or color, the diversity of feathers of seduction found among the birds-of-paradise is astonishing. They highlight just how incredibly versatile a structure the avian feather has evolved to be. —ES

WAHNES'S PAROTIA

SUPERB BIRD-OF-PARADISE

BLUE BIRD-OF-PARADISE

STRUCTURAL COLORS

These blue and green tones are all created by the selective reflection and refraction of light within the microscopic feather or skin structure.

VICTORIA'S RIFLEBIRD

KING BIRD-OF-PARADISE

GREATER BIR
OF-PARADISE

PIGMENTED COLORS

These yellow, orange, red, and chestnut colors are created by pigments that selectively absorb and reflect certain wavelengths.

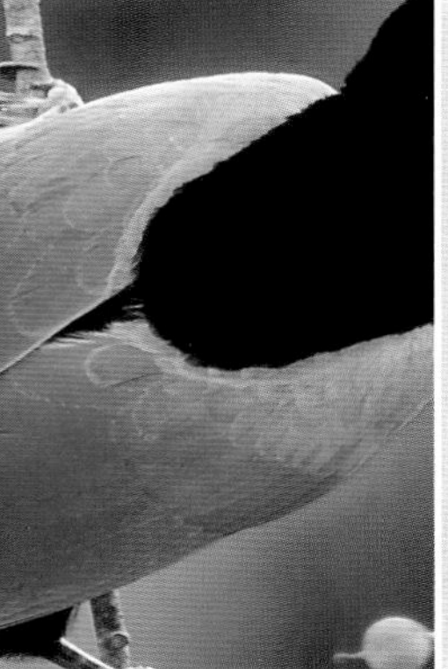

STANDARDWING BIRD-OF-PARADISE

RIBBON-TAILED ASTRAPIA

SPLENDID ASTRAPIA

WILSON'S BIRD-OF-PARADISE

RAGGIANA BIRD-OF-PARADISE

WILSON'S BIRD-OF-PARADISE

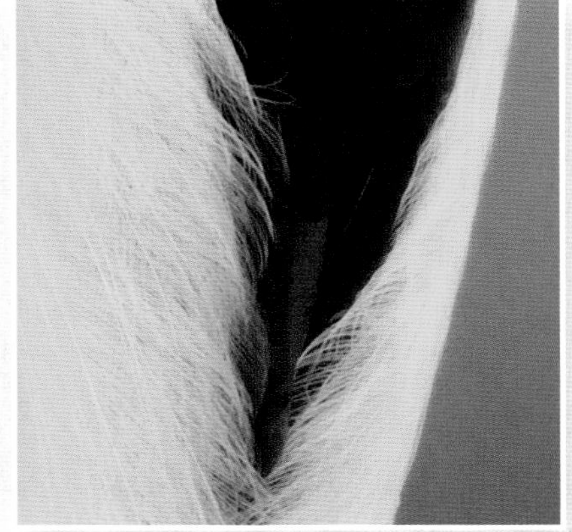

TWELVE-WIRED BIRD-OF-PARADISE

GREATER BIRD-OF-PARADISE

Payakona, Mount Hagen area, 15 September 2004

The feathers of more than 50 individual male birds-of-paradise are visible on the heads of these dancers performing at a compensation ceremony. In this ritual and many others, birds-of-paradise are important traditional symbols among New Guinean highland cultures.

People and Plumes

While we don't know exactly how long birds-of-paradise have been used as objects of material wealth and adornment, we do know people have inhabited parts of New Guinea for tens of thousands of years. In all likelihood, the relationship among people and birds-of-paradise is as old as the arrival of people themselves, since these birds feature significantly in the mythologies of many cultures.

To some indigenous people, birds-of-paradise represent the spirits of dead ancestors. Others feel they are the physical embodiment of mythological forest-dwelling beings. The courtship displays of many species have inspired a multitude of traditional dances, theater, and other types of art. To this day, the feather ornaments used by birds-of-paradise to attract mates are also used by people for the same purpose. Bird-of-paradise plumes have long been exchanged in the "bride wealth" payments given among families for the betrothal of daughters. In some places, birds-of-paradise serve as characters in marvelous tales. Like the fables of Aesop in the Western tradition, they are used to teach life lessons to young people so that they may learn to thrive among the many challenges they will face in a life spent in a remote tropical wilderness.

Without a doubt, the most astonishing connection between the people of New Guinea and the birds-of-paradise comes from the use of plumes as body decoration during "sing-sings"—the traditional New Guinean cultural gatherings, which involve song and dance. Traditionally used for ritualized warfare, courtship, and nearly every form of social interaction in between, sing-sings are still a major part of New Guinean culture, especially among highland populations.

On seeing a sing-sing for the first time in the highlands of Papua New Guinea in 1955, David Attenborough writes: "It was one of the most spectacular and barbaric sights I have ever seen. I made a rough calculation. There were over five hundred beplumed dancers. Between them, they must have killed at least ten thousand birds-of-paradise to adorn themselves for this ceremony." Today the numbers of both individual dancers and individual birds-of-paradise used as ornamentation are smaller. Traditional cultures are giving way to modern influences, and it's becoming common to see a peacock or rooster feather in a headdress at a sing-sing alongside the tail feathers of an astrapia.

Payakona, Mount Hagen area, 15 September 2004

Attractive to both female birds and humans, the two incredibly shaped feathers from the head of a male King of Saxony Bird-of-Paradise have been fed through the pierced septum of this tribal leader as part of his dress for a traditional ceremony. *(left)*

Mount Hagen Cultural Show, 21 August 2005

Huli Wigmen from the Tari area prepare to dance with the long white feathers of the Ribbon-tailed Astrapia and plumes from several other birds-of-paradise in their headdresses. Ceremonial use of feathers is common throughout the New Guinea highlands. *(right)*

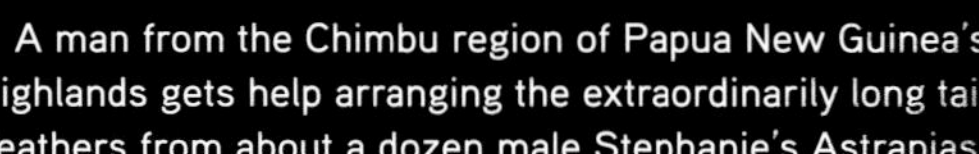

Mount Hagen Cultural Show, 20 August 2005

A man from the Chimbu region of Papua New Guinea's highlands gets help arranging the extraordinarily long tail feathers from about a dozen male Stephanie's Astrapias.

At the dance the Papuan will spend hours "dressing," then suddenly he will burst forth under a shimmering crown fit for an emperor. Sometimes the crown contains the plumes of a dozen or more males of half a dozen species and when fifty or so men get together in a dance the swaying plumage makes them appear to be on fire.

—E. THOMAS GILLIARD, *BIRDS OF PARADISE AND BOWER BIRDS*, 1969

A CLOSER LOOK

Displaying male parotias are masters of transformation. They shape-shift from their normal birdlike form into a spectacle that looks like either a bird wearing a ballerina's tutu or a very unbirdlike dancing ovoid shape.

NEW DISCOVERIES: *Parotia*

SECRETS OF THE COURT DANCERS

I started researching the species in genus *Parotia* for a very practical reason: Their courtship displays happen on the ground, which makes them much easier to witness than those in the trees! Their terrestrial display areas, which range from the size of a welcome mat to a throw rug, are relatively easy to locate in the forest—once you know the right places to look. Males make their ground "courts" along ridge tops, usually broad ones that aren't too steep. Walking ridges and listening for their raucous squawks is usually a pleasant experience. But finding a ridge that contains courts, when you are trekking among a network of ridges, is the tricky part. Even if you hear a parotia making a commotion on the next ridge over, it usually means making a slog down a steep slope, crossing a stream, and climbing up the other side. That simple traverse can be exhausting.

But once on the right ridge, finding the court isn't too difficult. Parotia courts are meticulously groomed surfaces within a sea of lush forest vegetation. The ground is devoid of the growth that usually proliferates wildly on the forest floor. Next to the court often is a pile of leaves, in various stages of decay, which have been tossed aside by the male while maintaining his stage. To the experienced eye, these factors add up to small but noticeable differences in the forest near a parotia court.

Displaying male parotias are masters of transformation. They shape-shift from their normal birdlike form into a spectacle that, depending on your vantage point, looks like either a bird wearing a ballerina's tutu or a very unbirdlike dancing ovoid shape (see pages 126-128). During my years of field research, I've found that parotia display repertoires are very complex and include a wide range of behaviors beyond the quintessential ballerina display. For example, I discovered that Carola's Parotia males not only *remove* leaves from their display courts but also *bring* other leaves to their courts to use as "props" during certain displays. Males hold small leaves in their bills while displaying to a female. This is the only known case of a bird-of-paradise using a prop.

Some of the behaviors among different species, at first glance, appear totally unique, yet are actually very similar and have evolutionary origins in the same underlying behaviors. What makes them unique are the ways in which sexual selection has elaborated them. However, break them into their component parts and it's clear that they are derived from the same ancestral behaviors. A good example of this can be found

This Carola's Parotia male holds a yellow leaf in his bill during his display, an interesting example of a bird using a "prop" for added ornamentation. Koko-o Crater Mountain, 24 September 2005

Caught in midair, a male Western Parotia performs the court-clearing dance, a display derived from the simpler behavior of cleaning the court *(view video 55310 at macaulaylibrary.org)*. Lower Syoubri, Arfak Mountains, 2 December 2004

During the "swaying bounce" display, a male Carola's Parotia near Crater Mountain flutters his wings and wobbles up and down and side to side. Koko-o Crater Mountain, 25 September 2005

In a slow exposure of the same behavior, the white and bronze feathers on his head trace his repeated figure-eight motion *(view video 56733 at macaulaylibrary.org)*. Koko-o Crater Mountain, 24 September 2005

in a display of the Western and Carola's Parotias, which are embellished versions of the simpler court-clearing behavior. All parotia males remove leaves and other forest debris from their courts. They systematically hop around their stage, and when they find a leaf, they bend down headfirst to pick it up with their bill and, with a flick of the neck, toss it aside. The sequence of the cleaning goes like this: hop forward, bend down so the bill nearly touches the ground, grab the leaf, raise up while flicking back the head to toss it, and repeat. In the Western and Carola's Parotias, this sequence has been elaborated upon and ritualized through sexual selection in similar but distinct ways.

The Western Parotia male performs what can clearly be identified as a court-clearing display. The display looks similar to typical court clearing, but its movements are stereotyped and it takes place at a particular point during the courtship sequence. In other words, it has become ritualized. In this version, the male hops into the central part of the court, bends down as if to pick up a leaf, but only mimes pecking at the ground, and then lifts up, but with more vigor than in typical court-clearing behavior, while flicking his head back and hopping off the ground and to one side, while also flicking his wings open and closed. The sequence of ***bend down, mime-peck, hop to side, wing-flick*** is repeated several times before moving on to another display.

In Carola's Parotia, the display derived from court clearing is called the "swaying bounce." The male stands in the middle of his court, with his back to the females above on the horizontal perch, and "bounces" up and down by raising and lowering his body with his legs while also swaying from side to side in a rhythmic figure-eight pattern. During the swaying bounce, his wings are opened to the sides and fluttered.

In fact, if you could take the hops to the side and wing-flicks of the Western Parotia, speed them up, and loop them back and forth, the result would appear surprisingly similar to the swaying and bouncing of the Carola's Parotia: up-and-down, side-to-side motion, with wings open and back toward the observing female above.

Another piece of the puzzle strengthens the evidence that these two seemingly unrelated behaviors are actually one and the same, both being evolutionary elaborations of the simpler court-clearing behavior. As the Carola's Parotia is preparing to begin the swaying bounce, he hops into the central part of his court, lowers his head to the ground, and mimes picking up a leaf or other debris with his bill. Just as he starts to enter the full-blown swaying-bounce behavior, he raises up and tosses his head back in a distinctive way, and then turns his back away from the female and starts to sway in the figure-eight motion and bounce up and down. In that brief instant at the very beginning of the sequence are the evolutionary foundations of the much elaborated courtship display revealed. But once you see it with a trained eye, you cannot help drawing the ultimate conclusion: the swaying bounce is nothing but glorified court clearing. And that is pretty amazing! —ES

Slowly moving closer to the female, an adult male Carola's Parotia engages in a head-tilting display, turning his head while fluttering the feathers on his chin. Koko-o Crater Mountain, 26 September 2005

Leaning in toward the female, this adult male Western Parotia performs the "horizontal perch sidle"—a display done on a perch above the court. Lower Syoubri, Arfak Mountains, 6 December 2004

This typical patch of forest floor from a midmountain forest in the Arfak Mountains is directly adjacent to the parotia display court shown below.
Lower Syoubri, Arfak Mountains,
6 December 2004

The well-groomed court of a Western Parotia stands out among the chaos of the forest. This court was maintained for at least six years, from 2004 to 2010.
Lower Syoubri, Arfak Mountains,
6 December 2004

WESTERN PAROTIA ~ *Parotia sefilata*

Lower Syoubri, Arfak Mountains, 9 August 2009

Not to be confused with wings, the velvet black "skirt" of this male is made from special upper breast and flank feathers that evolved for the sole purpose of transforming his body shape.

Masters of Transformation

Of all the bizarre behaviors among birds-of-paradise, the most incredible are those of the species in which the males change, in the blink of an eye, from a bird into an almost otherworldly being. At first glance, these masters of transformation may not appear as stunning as the more colorful species, but once you understand what happens as they adopt their nonavian shapes, they become quite astounding indeed.

The shapes adopted by the shape-shifting birds are without precedent. Whatever image of a bird we harbor in our mind's eye is completely inadequate. When people see a shape-shifting display for the first time, they know that the performance is bizarre, but few comprehend exactly what they are seeing. That's because the shapes are put together from a combination of specialized feathers and behaviors that other birds don't have. When seeing one of the shapes emerge during a display, our cognitive faculties assume, from previous experience with every bird we've ever seen, that the unusual ovoid shape must be made out of the parts, like wings, that we know all birds have. And in some cases, the wings *are* being used. But in other cases, our minds deceive us, because once again the truth is stranger than fiction.

The unfamiliar ovoid shapes we see among the shape-shifting species are not actually individual structures at all. They are not a fabric like a cape, a hood, or a skirt, but rather are composite "meta-structures" made from the precise alignment of dozens, sometimes hundreds, of independently movable parts. In some cases, the color patches that emerge on these apparent structures don't even exist: The bird has coordinated the alignment of many individual feathers with specific body movements that create the illusion. A shining example of this sleight of hand is the shimmering blue line that emerges along the periphery of the ovoid shape produced by a male Black Sicklebill. Another is produced by the male Superb Bird-of-Paradise, which displays a metallic blue "smiley face." Both are astonishing to behold, leaving the onlooker speechless. These are indeed masters of illusion.

MAGNIFICENT RIFLEBIRD ~ *Ptiloris magnificus*

Oransbari, Bird's Head Peninsula, 21 August 2009

Mouth agape while making his loud wolf whistle from the horizontal vine used for courtship, this adult male is quite magnificent, as his species name suggests, but nonetheless distinctly birdlike in appearance. *(left)*

MAGNIFICENT RIFLEBIRD ~ *Ptiloris magnificus*

Oransbari, Bird's Head Peninsula, 21 August 2009

In an instant, upon the approach of a female, the same male *(shown at left)* transforms himself into a fantastic black ovoid shape. His head, now a mere blur of brilliant blue, whips vigorously from side to side as the ovoid form moves up and down to the rhythm of this impressive display *(view video 55444 at macaulaylibrary.org)*. *(right)*

BLACK SICKLEBILL ~ *Epimachus fastosus*
Bog Camp, Foja Mountains, 23 June 2007

The ovoid form of this adult male is a prime example of how truth is stranger than fiction. For two centuries, illustrators speculated how the male's bizarre "epaulette" feathers might be used. Even in their wildest imagination, no one dared propose anything as incredible as what sexual selection actually devised.

Birds of paradise are famous also for their strange and beautiful dance movements which frequently make the males appear startling and grotesque and more ornament-like than bird-like.

—E. THOMAS GILLIARD, *BIRDS OF PARADISE AND BOWER BIRDS*, 1969

WAHNES'S PAROTIA ~ *Parotia wahnesi*
Upper Satop, Huon Peninsula, 31 July 2007

As an adult male Wahnes's Parotia bows deeply to start his performance, his long tail makes this relatively minor courtship detail quite impressive. But what happens in the instant the male lifts up from the bow is the most mind-blowing *(view video 56818 at macaulaylibrary.org)*. *(above)*

WAHNES'S PAROTIA ~ *Parotia wahnesi*
Upper Satop, Huon Peninsula, 31 July 2007

In a blink of the eye, he tucks his head to one side, gathers himself up as if taking a deep breath, and bursts forth with flank-plumes expanded around his body. Once he is upright, the long, flag-tipped, wirelike feathers behind each eye push forward to punctuate the already curious performance. *(right)*

I have listened to these birds calling from nearby trees for days on end, waiting for a chance to photograph a display.

A male Superb Bird-of-Paradise spreads his cape in a brief display. Unfortunately, he faced away from Tim's camera during the entire episode. Tigibi, Tari area, 29 October 2011

THE CHALLENGING SHAPE-SHIFTERS

MY NEMESIS—THE SUPERB BIRD-OF-PARADISE

Ed's video was jaw-dropping. This had to be one of the most incredible behaviors in the animal world. He showed it to me during one of our first meetings in 2004, before we ever traveled to New Guinea together. Ed had been doing fieldwork in New Guinea for several years already, especially studying *Parotia* species for his Ph.D. One day during a trip in the Adelbert Range, he had heard a Superb Bird-of-Paradise displaying and found its display site—a fallen tree, as is typical for this species. Setting up a blind, he had filmed the incredible transformation the male makes when a female arrives and he raises his cape (see the series of video frames on the opposite page).

I knew I absolutely had to shoot still photographs of this behavior. As far as I knew, it had never been photographed. It became one of my top priorities, and I have been trying ever since. That's where the "nemesis" part comes in.

We have located displaying Superb males in several locations on different expeditions, which is not surprising because it is a widespread species on New Guinea. We have hired dozens of scouts to monitor them and try to identify a male that displayed predictably. I have built blinds and spent literally weeks in them. But I have yet to photograph a male displaying—at least while facing toward me.

I have listened to them calling from nearby trees for days on end, waiting for a chance to photograph a display. I have seen them flutter down to their display log in front of me and call enthusiastically, trying to lure a female. They have twitched their cape as if they were about to open it, causing sudden spikes in my blood pressure. I have photographed a male, cape fully spread, apparently with a female in his sights—but he was facing away from me the entire time.

I even devoted the last three weeks of fieldwork in 2011 to capturing an image of a displaying Superb Bird-of-Paradise. Finally, we located a perfect male Superb who was very active. The problem was, he had many display sites around his territory and refused to use the same one twice. During this last effort, our Huli guide was so frustrated with the male Superb that he told me, "If this male does not display for you tomorrow"—my last day—"I am going to shoot him with my bow and arrow." I certainly hope he didn't do that.

THE ELUSIVE BLACK SICKLEBILL

Luck came calling on my very first morning in the remote Foja Mountains of Papua, Indonesia. A helicopter had dropped us off the day before in a muddy bog, high in these rugged mountains.

Tim spent six days in a blind at this display log and photographed only one visit by the male Superb Bird-of-Paradise shown here—but no display. Ambua, Tari area, 3 December 2010

A video sequence recorded by Ed in the Adelbert Mountains in November 2001 shows the shape-shifting display of the Superb Bird-of-Paradise *(view video 58003 at macaulaylibrary.org)*.

A Superb Bird-of-Paradise male calls from a display log, trying to lure in a female.
Tim spent over two weeks trying to photograph this particular male displaying without success.
Tigibi, Tari area, 5 November 2011

We were probably a two-week walk from the nearest village (even if there had been a trail). Dawn was just breaking, and I was on the hill above our camp when the sound started—the unmistakable, very loud whiplike *quink-quink* call of a Black Sicklebill. Not one but two different males were calling very close by in the misty, moss-cloaked forest. What luck! They were near enough that I might be able to home in on them while they were displaying. I crept through the forest, trying to get a glimpse of the birds. First I spotted one, and then the other—long black shapes in the gloom. They were both perched on broken-off trees, the classic Black Sicklebill display site. Not only were both calling, but they were also performing their incredible shape-shifting display, transforming into shapes so arresting that they seemed more like a movie alien than a bird.

I knew that this species of bird-of-paradise was poorly documented. A full display in the wild had never been recorded. Few people had ever witnessed this display anywhere, and now, standing unnoticed in this vast wilderness, I felt the thrill of seeing a true marvel of nature for the first time. Knowing how remote my location was, I felt confident I was the first person ever to see these two birds display here. Not only that, I felt sure Black Sicklebills had never been reported to display in pairs, so I was observing something totally new.

Over the coming days, I documented as best I could, in the often heavy mist of the Foja Mountains, the dawn displays of the Black Sicklebills. On one occasion, I even filmed a female visit, which had never been previously documented. Who would have guessed that she came right to the display pole and looked the male over from centimeters away (see video frames at right)? By the time the helicopter came back to pick us up, I had amassed the largest collection of Black Sicklebill stills and video ever made. That information constituted a significant contribution to the scientific understanding of bird-of-paradise behavior.

Alfred Russel Wallace wrote the following upon leaving his field site in the Aru Islands. I think it perfectly captures the feeling Ed and I have about working in New Guinea on birds-of-paradise.

> *I fully intended to come back; and had I known that circumstances would have prevented my doing so, should have felt some sorrow in leaving a place where I had first seen so many rare and beautiful living things, and had so fully enjoyed the pleasure which fills the heart of the naturalist when he is so fortunate as to discover a district hitherto unexplored, and where every day brings forth new and unexpected treasures.*

There is so much more to learn. We'll be back.

—TL

These two frames from a video sequence show something never seen before: a female Black Sicklebill visiting the pole of a displaying male *(view video 56281 at macaulaylibrary.org)*. Bog Camp, Foja Mountains, 17 June 2007

An adult male Black Sicklebill atop his display pole in the Arfak Mountains. Note how the epaulette feathers are exposed at his sides but not erected as they would be in display. Upper Syoubri, Arfak Mountains, 12 August 2009

These two views of a brief display by the same male shown in the photo at left come from a video sequence Ed recorded while Tim shot stills from a blind nearby. Note that the male's wings are held against his back, while he raises his epaulette feathers around his head. Upper Syoubri, Arfak Mountains, 12 August 2009

KING OF SAXONY BIRD-OF-PARADISE ~ *Pteridophora alberti*
Mid Tari Gap, 25 October 2011

As the morning mist clears, an adult male emerges to sit on his favorite calling perch. The King of Saxony Bird-of-Paradise, an iconic bird of the New Guinea highlands, is not only a marvel of evolution but also a symbol of the majesty of wild places and our need to explore and steward them.

The Birds-of-Paradise Project by the Numbers

Duration of project in years	8
Number of expeditions	18
Number of field sites visited	51
Days on expeditions	544
Commercial flights taken	200
Bush plane/helicopter flights taken	33
Number of aircraft we flew in that crashed (later)	2
Boat trips taken	58
Number of times adrift at sea in broken-down boats	2
Photography blinds/hides built	109
Man-hours spent in blinds	2,006
Number of hours Ed spent in a blind to film a Riflebird display	80
Number of times Ed saw a Riflebird display (for 90 seconds)	1
Tree climbs made for photography	146
Height above ground of Tim's highest canopy blind (in meters)	50
Number of times an appendix burst in a remote camp (Ed's)	1
Number of days to reach surgery with a burst appendix	5
Audio and video recordings archived at the Lab of Ornithology	2,256
Photographs brought back (after deleting rejects)	39,568

Species Atlas

LC = Least Concern

NT = Near Threatened

VU = Vulnerable

ELEVATION RANGE

5,000 / 16,404; 4,500 / 14,764; 4,000 / 13,123; 3,500 / 11,483; 3,000 / 9,483; 2,500 / 8,202; 2,000 / 6,562; 1,500 / 4,921; 1,000 / 3,281; 500 / 1,640; 0 m / 0 ft — sea level

MAP KEY

species range

8○ expedition site and number

see page 19 for complete list of expedition sites

Pigeon

This last section contains a brief overview of the 39 birds-of-paradise we studied during the eight years of the project. Many have been discussed in the preceding pages, but here they are systematically organized for easy reference. What follows is a half-page account per species with seven components included for each: (1) additional photos, (2) a short essay, (3) a range map, (4) an elevation range graphic, (5) an illustration showing relative size, (6) a brief voice description, and (7) conservation status.

You should be aware of several variations as you use the atlas. Our classification sequence departs from previous works because we chose to follow the most up-to-date "family tree" (phylogeny) from the scientific literature. This isn't the last word by any means, but we feel it's an improvement. The naming conventions we follow are all used in other sources, but we didn't follow any one source strictly, as none seemed entirely adequate. Yet we tried to be as conservative as possible.

Photographs were selected to show a wider range of natural history information, including (when available) both sexes, alternative plumages, and noteworthy behaviors. The essays briefly introduce each species and convey a few aspects we find interesting about each; those desiring more information should consult the excellent monograph *The Birds of Paradise* by Clifford Frith and Bruce Beehler or the more recent accounts in the *Handbook of Birds of the World* series.

The range maps depict the global distribution of each species and are modified from the master's thesis of Papua New Guinean biologist Leo Legra. The maps are based on ecological niche models that use point-locality and environmental data to model the species' range in a much more informed way than predictive range maps usually do. We also provide a graphic to show for each species a general elevational range—a major factor influencing how different species of birds-of-paradise are distributed.

To convey relative size, we provide a visual comparison to a pigeon—a species that most people around the world are familiar with. The voice description covers only the species' primary vocalization, which is almost always made by the male, and rarely includes information about vocal variations (of which there are many). Finally, the species conservation status posts what is listed on the most current IUCN Red List (Version 3.1).

Pair staying together. One preening its wing. Site 28, 15 Jul 08

PARADISE-CROW ~ *Lycocorax pyrrhopterus*

Is it a crow or a bird-of-paradise? While the common name of this species leads to confusion, it also underscores just how crowlike the birds-of-paradise really are. The Paradise-crow bears the most resemblance to the crowlike common ancestor that gave rise to the birds-of-paradise.

But if it's a bird-of-paradise, why is it called a crow at all? The answer dates to 1851, when Charles Lucien Bonaparte, a French ornithologist and nephew of Napoleon, first described the species for science. At the time, he placed it in the genus *Corvus*, which includes the 40 or so species of crows (and ravens) in the family Corvidae. But two years later, Bonaparte reevaluated the evidence and decided that this species wasn't a crow after all, but was instead a very crow-like bird-of-paradise. As a result, he created the genus *Lycocorax* (which means "wolf raven" in Latin). Bonaparte was not the only scientist who was uncertain. In 1869, Alfred Russel Wallace thought this species was a starling, and in 1886, F. H. H. Guillemard believed it to be a "miniature crow approaching the Paradisaeidae"—but a crow nonetheless. However, current evidence supports classifying the Paradise-crow as a bird-of-paradise. A sister form to the manucodes, it is the earliest known offshoot from the bird-of-paradise family tree.

The Paradise-crow is the only member of its genus. Because it is a shy forest dweller confined to the northern Moluccas, it is poorly known. Other than the brown color of the wings, the species is very crowlike in appearance, possessing no iridescence or other ornate features of other birds-of-paradise. Males and females resemble each other in appearance, and although little is known of the Paradise-crow's breeding behavior, it is presumed to be monogamous.

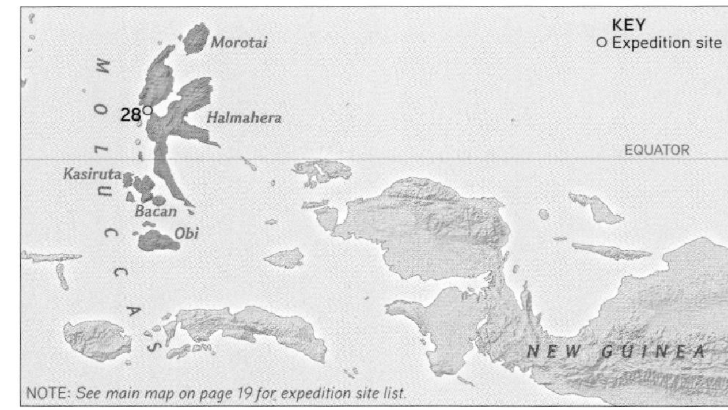

NOTE: *See main map on page 19 for expedition site list.*

ELEVATION RANGE

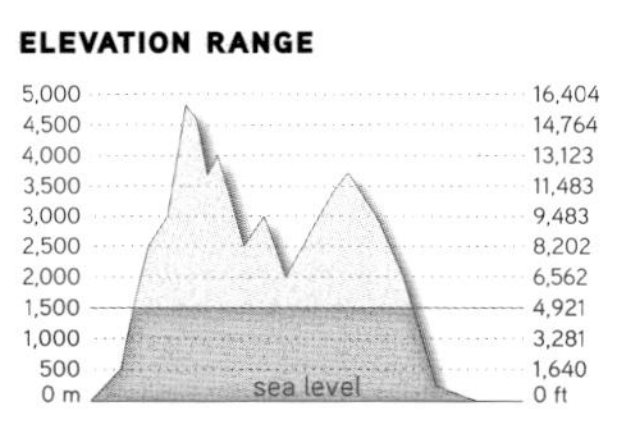

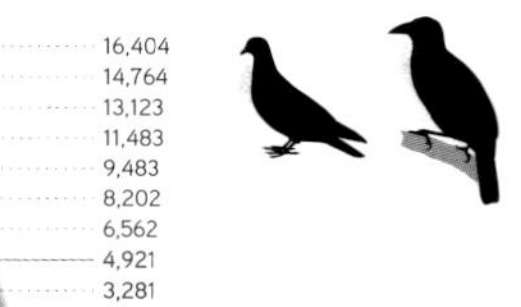

CONSERVATION STATUS LC

VOICE Harsh barking *wunk* notes, loud double *kek-kek* clicks, and a resounding trumpet-like *yeung*

Adult eating fruit. Site 4, 1 Oct 11

TRUMPET MANUCODE ~ *Phonygammus keraudrenii*

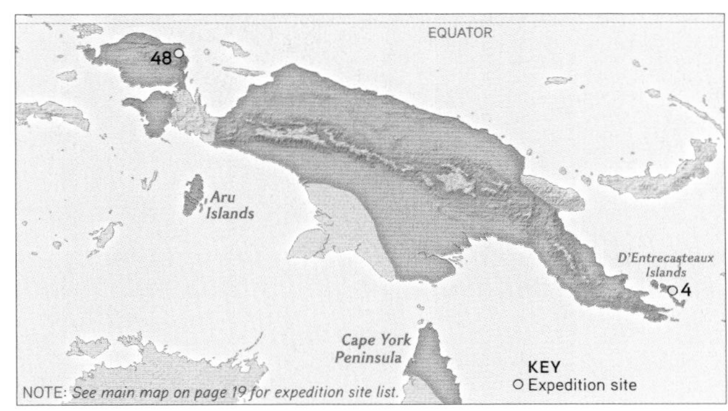

With a common name reflecting its far-carrying "trumpet-like" vocalization, the Trumpet Manucode is arguably the most distinctive of the manucodes, which is also why it is in its own genus. Although noticeably smaller than the others, this species is most easily distinguished by its somewhat shaggy appearance, which comes from its mane of elongated hackle-like feathers on the neck, nape, and upper breast. It also has prominent ear tufts, which extend from above the eyes and rear crown and add to the overall shaggy look of the upper body.

This species has the distinction of being the most widespread of all the birds-of-paradise. Its range spans all major regions of mainland New Guinea, the Aru Islands southwest of New Guinea, the D'Entrecasteaux Islands in the east, and the Cape York Peninsula of Australia in the south. The Trumpet Manucode also has the broadest elevation range of any bird-of-paradise. It extends from lowland rain forests near sea level to mountain cloud forests above 2,000 meters (6,600 feet) in elevation. Like most birds-of-paradise, its diet primarily consists of fruit, but some evidence suggests that this species consumes a higher proportion of figs than the others. Interestingly, the Trumpet Manucode may even provide its nestlings largely with figs—a rare practice among birds.

The unusually elongated trachea (windpipe) found in most manucodes is particularly remarkable in the Trumpet Manucode. Males have up to six concentric loops nestled between the skin and breast muscles. This specialized feature undoubtedly plays a significant role in the species' distinctive far-reaching vocalizations. It has likely evolved to promote sexual selection, since females have no tracheal loops.

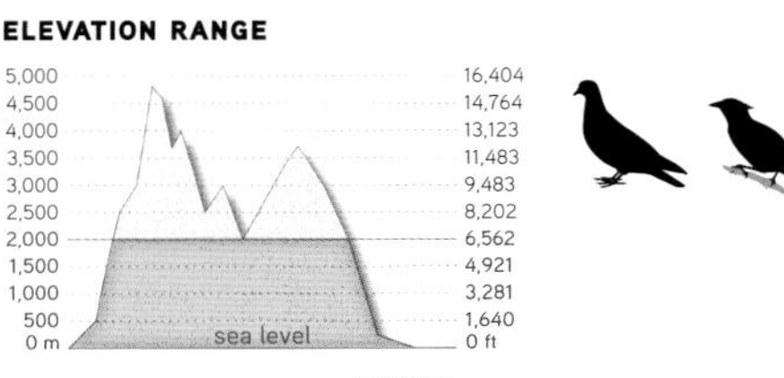

CONSERVATION STATUS LC

VOICE Trumpeting *grrwaawk, gyowlp,* or *graalp* growls

Adult sitting on nest. Site 5, 25 Sep 04

Nest with two chicks being fed. Site 5, 25 Sep 04

CURL-CRESTED MANUCODE ~ *Manucodia comrii*

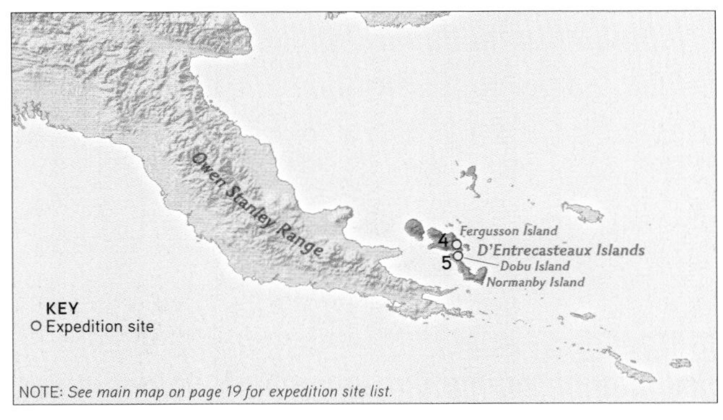

The largest and most conspicuous species of manucode, the Curl-crested Manucode is one of the more commonly encountered birds in the forested parts of the D'Entrecasteaux Islands. From afar, it appears rather crowlike because of its blue-black color and relatively large size. However at close range, its distinctive plumage features emerge. As its common English name implies, the top of the head contains a crest of oddly curled and frizzled feathers. When erected, these make the bird look like it is wearing a top hat. The feathers of the neck, breast, and back are also curled and frizzled, which give the body a uniquely plush and spangled texture. The unusual tail has two central feathers that are twisted so the undersides face inward to form an A-shape while the other tail feathers become the crossbar of the A. The overall appearance of the tail is vertically broadened and wedge shaped, similar to that of a Great-tailed Grackle's (*Quisicalus mexicanus*).

But you can't readily see what may be the most impressive feature of the Curl-crested Manucode: the looped subcutaneous trachea of adult males. It can extend the full length of the torso, loop around the abdomen, and return back up to the throat. Because its trachea is so unusually long, the vocalizations of this species are quite extraordinary. The low hollow sounds are resonant, conspicuous, and far carrying. In the forests of Fergusson Island, we could routinely hear four individuals (likely two pairs) vocalizing back and forth in response to one another throughout most of the day. On one memorable occasion, we were awakened on a brightly moonlit night to the haunting sounds of what seemed to be half a dozen individuals duetting back and forth in the forest near our camp.

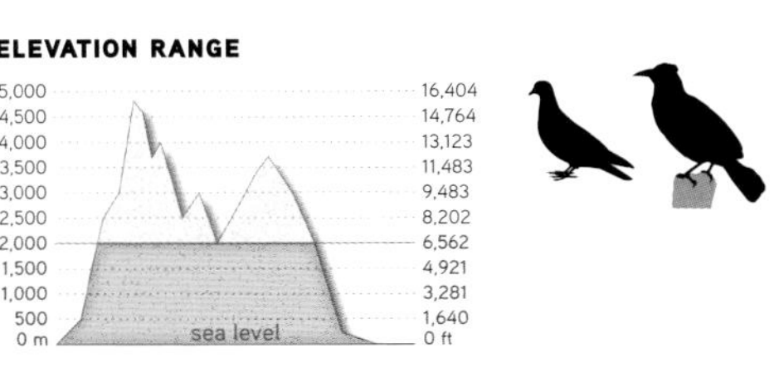

CONSERVATION STATUS LC

VOICE A hollow resonant low-pitched and slightly descending whistled *kwo-kwo-kwo-kwo-kwo-oo-oo-oo*

SEE ALSO P 42

Eating a fig. Site 48, 14 Jun 11

Adult. Site 48, 15 Jun 11

Adult. Site 48, 15 Jun 11

CRINKLE-COLLARED MANUCODE ~ *Manucodia chalybatus*

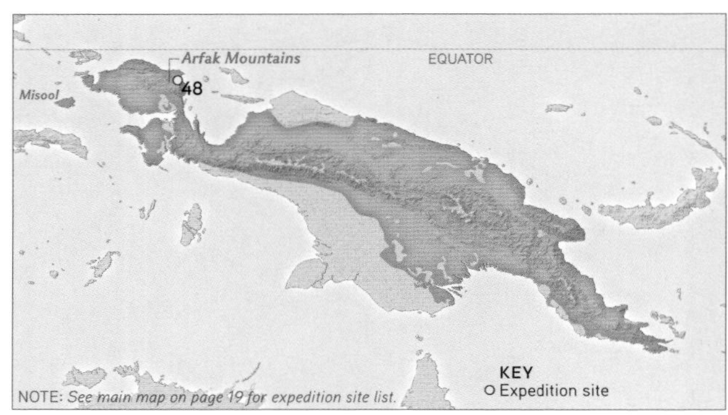

The Crinkle-collared Manucode is arguably the easiest to pick out of the three "difficult to distinguish" manucodes (the others being the Glossy-mantled and Jobi Manucodes). While it is found in the lowlands in some parts of its range (e.g., the island of Misool) and may occasionally be found in lowland forest near hills, it is primarily a hill forest species, roughly above 100 and below 1,000 meters (330–3,300 feet) elevation. Throughout most of its broad range, it's likely to be found alongside a Trumpet Manucode (as Tim photographed in a fig tree in the Arfak Mountains), which is easier to identify. The Crinkle-collared Manucode has a visibly longer tail and generally appears more elongated than the other two tough-to-differentiate manucodes. In addition, it possesses a prominent eyebrow tuft, which can be quite visible at times. Unfortunately, the crinkled collar that gives it its name is rarely seen from afar.

The behavior of this species is slightly better known than other manucodes, perhaps because it is more frequently encountered. When moving about the forest, it tends to make jerky side-to-side body movements that make it appear wary (which may be the case). When vocalizing, it performs a simple display similar to that of the Glossy-mantled Manucode: It expands the upper body plumage and lifts and partially spreads the wings and tail with each note. The Crinkle-collared Manucode also favors eating figs. One study found that 93 percent of the birds in 59 observed feedings were consuming figs.

ELEVATION RANGE

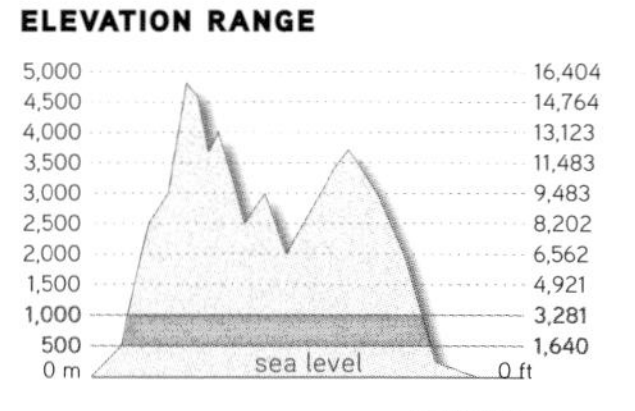

CONSERVATION STATUS

VOICE Slow series of hollow whistled *hoo* or *who-oo* notes and a variable *tuk, chenk,* or *chook* note

---

Adult. Site 49, 21 Jun 11

Adult. Site 49, 21 Jun 11

Adult. Site 49, 21 Jun 11

JOBI MANUCODE ~ *Manucodia jobiensis*

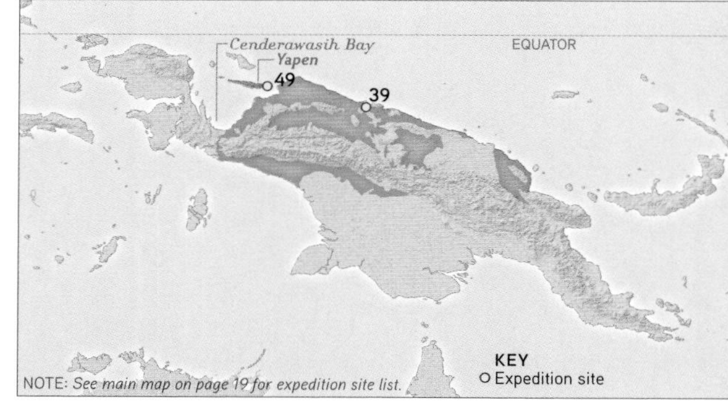

This species derives its name from the island of Jobi, which is an outdated name for the island now known as Yapen in Cenderawasih Bay of western New Guinea. Importantly, it is the only manucode species that inhabits the forests of Yapen—if you see a manucode there, it must be a Jobi. This fact was critical for our project, because the Jobi Manucode is a secretive and poorly known species that is difficult to distinguish from its two all-but-identical sister species: the Glossy-mantled and Crinkle-collared Manucodes. Equally as bedeviling, for most of its range throughout the northern lowlands of mainland New Guinea the Jobi Manucode coexists with one or both of these more widespread species. From museum specimens, we know that the Jobi is, on average, slightly smaller with a somewhat shorter trail and a blunter bill, but these traits are difficult to assess in the field. The Jobi does lack the defining feature of the Glossy-mantled Manucode—its glossy "mantle" or back—but this trait varies with age; younger birds of both species are a duller, sootier black.

Even with a good-quality photograph to scrutinize, it's very difficult to state positively that you've seen a Jobi Manucode. The most reliable way to distinguish the Jobi from the others is by voice, but the species isn't particularly vocal and, complicating matters, vocalizations of Jobi Manucodes are not well known. Thus, after several attempts to take photographs of the Jobi Manucode in mainland New Guinea, we definitively documented the species on a special expedition to "Jobi" island, where we could be sure!

ELEVATION RANGE

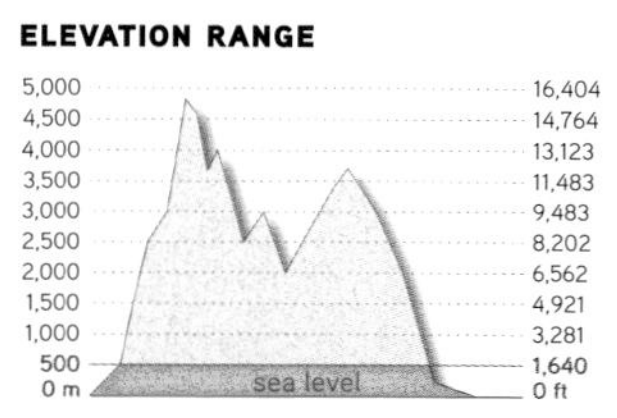

CONSERVATION STATUS LC

VOICE Little known; four to six hollow *hoo* notes and also a harsh *chig* sound

Subadult. Site 39, 30 Jun 10

Adult. Site 39, 14 Jun 10

Adult. Site 8, 10 Aug 05

GLOSSY-MANTLED MANUCODE ~ *Manucodia ater*

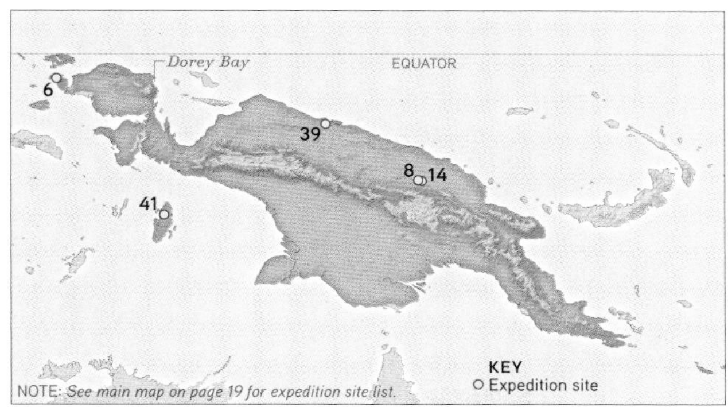

On July 26, 1824, the French ship *La Coquille* dropped anchor in Dorey Bay off the northwestern coast of mainland New Guinea near present-day Manokwari. René Primevère Lesson, the ship's surgeon, immediately went ashore to explore the surrounding forests and collect scientific specimens as part of his dual role as one of the ship's naturalists. Within the first few days, but entirely unknown to him at the time, Lesson became the first Westerner to see a living bird-of-paradise in the wild. The species he first encountered (and collected for science) wasn't the brilliantly plumed Lesser Bird-of-Paradise that made a lasting impression on him, but instead was a glossy, iridescent blue-black crowlike bird, which later became known as the Glossy-mantled Manucode.

The Glossy-mantled Manucode can barely be distinguished in the field from its closest relatives, the Jobi and Crinkle-collared Manucodes. Even at close range, its diagnostic features, like the smooth, glossy sheen on its upper back, are rarely discernible. The best tool for field identification is to hear its mournful drawn-out whistle, reminiscent of feedback from a microphone and speaker. Throughout most of its wide range, the Glossy-mantled Manucode co-occurs only with the Crinkle-collared Manucode, which is usually (but not always) found at higher elevations. Thus, in many parts of New Guinea, this species is the most commonly encountered manucode of the lowland forests. But without hearing its call, identification is still difficult.

The species performs a simple display while vocalizing. The (presumed) male expands the plumage of the upper body while leaning forward, rising on its legs, spreading the wings and tail slightly, and stretching its neck. The trachea is elongated, but to a lesser extent than other manucodes, with only a single short loop on the upper breast.

ELEVATION RANGE

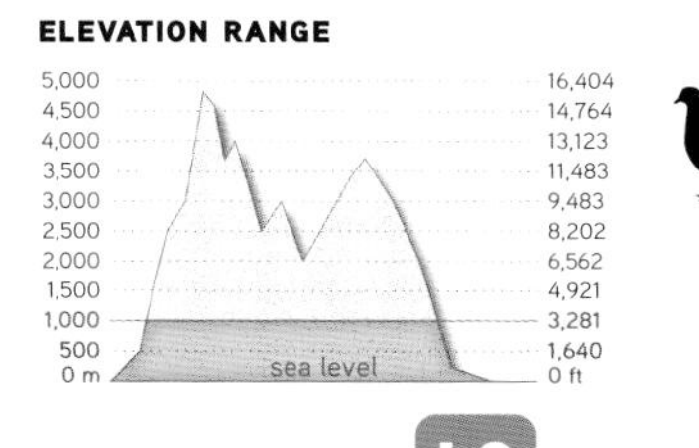

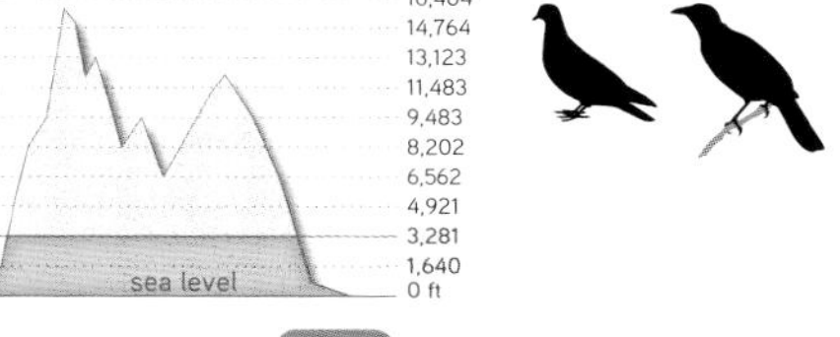

CONSERVATION STATUS LC

VOICE A pure-tone, high-pitched whistle and a soft *tuck* or *kwek* note

---

KING OF SAXONY BIRD-OF-PARADISE ~ *Pteridophora alberti*

Female plumage. Site 18, 27 Oct 11

Calling male. Site 18, 26 Oct 10

Displaying male. Site 18, 2 Dec 10

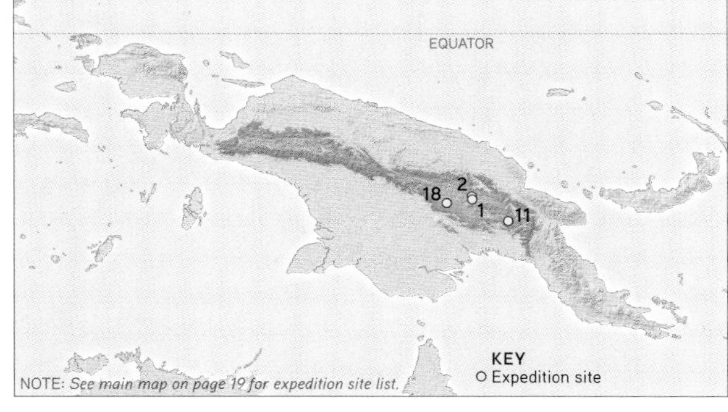

Without question, the King of Saxony is one of the most astonishing animals on the planet. Its ornamental head plumes, called "flags" or "head-wires," are unlike any other feather known to exist. These bizarre feathers can be twice as long as the male's body and their physical structure resembles feathers so little that some scientists were convinced that the first specimens were chimerical fakes contrived by unscrupulous collectors to fetch a high price. But they are indeed real, and they are prized as ornaments not only by female King of Saxony birds but also by humans and even another bird species. The feathers are worn in great numbers as part of the traditional dress of many New Guinean cultures, and they are also cherished bower decorations of the male Archbold's Bowerbird. Equally as bizarre as its ornamental head feathers is the male's voice, which sounds variously like pulsing electronic static, rushing water, or vigorous shaking of a baby rattle. It is an iconic sound of the central cordilleran cloud forests where this species is found.

But to really put this evolutionary oddity into perspective, consider this thought experiment: Imagine having two purely ornamental wirelike rods sticking out from your head near the temples. Now envision those rods as being more than twice as long as you are tall and studded along their entire length with plastic flaglike ornamental tabs. Then try to imagine being able to wave those head-wires in every direction, using nothing more than the muscles in the skin of your head (much like those used to wiggle your eyebrows). Now picture all of this while wearing a midlength black velvet cape and doing rhythmic deep knee bends to bounce up and down while standing on a tightrope. Sounds absolutely ridiculous, right? But the male King of Saxony Bird-of-Paradise during the breeding season performs this too-bizarre-to-be-true ritual nearly every day. And it is one of the most wonderful sights to behold in all of the natural world.

ELEVATION RANGE

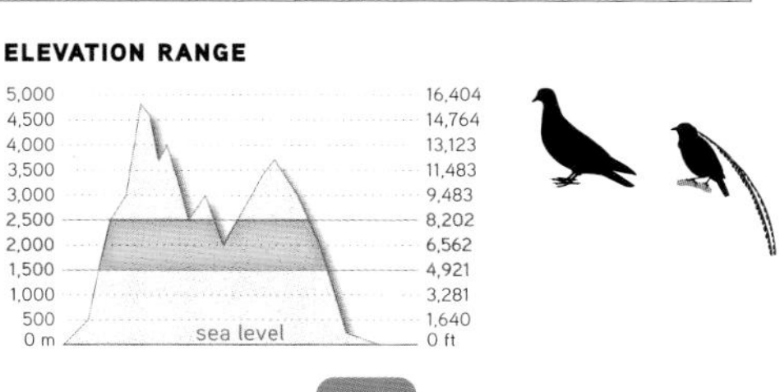

CONSERVATION STATUS

VOICE A pulsing nonavian static, rushing-water, or rattling sound with high-pitched screeching overtones and twittering subchatter

SEE ALSO PP 1, 156–157, 173, 198

Male and female. Site 11, 26 Sep 05

Male. Site 11, 25 Sep 05

Ballerina dance. Site 11, 25 Sep 05

SEE ALSO PP 120, 165, 182-183, 184

CAROLA'S PAROTIA ~ *Parotia carolae*

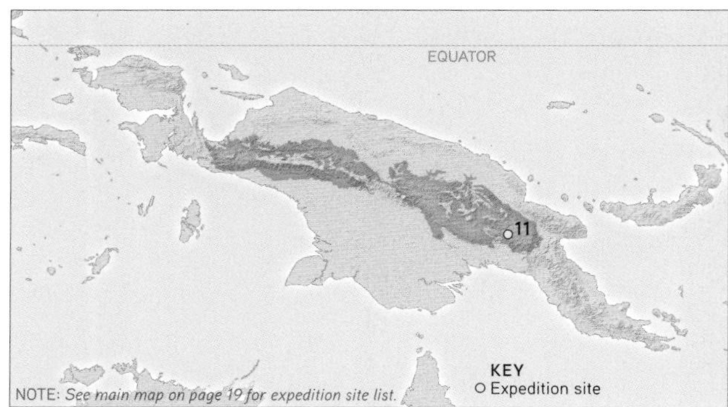

Carola's Parotia is named after Queen Carola of Saxony, wife of King Albert, for whom the King of Saxony Bird-of-Paradise is named. It is the smallest, most ornately plumed, and most widespread in the genus *Parotia*. As with all parotias, males build and maintain a specialized display area, or court, on the ground. These patches of forest floor, which are owned by individual males, are fastidiously cleared of all leaf litter and undergrowth to provide an open, well-lit stage for courtship performances. On these stages, which range in size from the equivalent of a welcome mat to a throw rug, males perform some of the most intricate and complex displays of all the birds-of-paradise. The males clear their court of as many sticks and leaves as possible, and they always select a spot with at least one branch that spans the court specifically for females to perch on while observing the courtship performance.

This species has several features and behaviors that set it apart. For one, they have bright yellow eyes, whereas most of the others' are cobalt blue (the other exception is the closely related Bronze Parotia). They also have unusual wispy throat feathers, or "whiskers," that are prominently wiggled during certain displays. Behaviorally, this species is the only bird-of-paradise known to use a prop—that is, an object brandished as part of their display. Males often bring yellowish leaves to their court and hold them in their bill while performing. Why they use a prop for display is uncertain, but it likely has to do with further exaggerating the effect of head movements. That might also explain why this species has ornamental facial markings when the other parotias don't (see "New Discoveries" on pages 182-185).

Vocally, this species also stands out from the others. Whereas most species emit harsh, raspy squawking sounds, this species gives various whistles and notes of pure tone.

ELEVATION RANGE

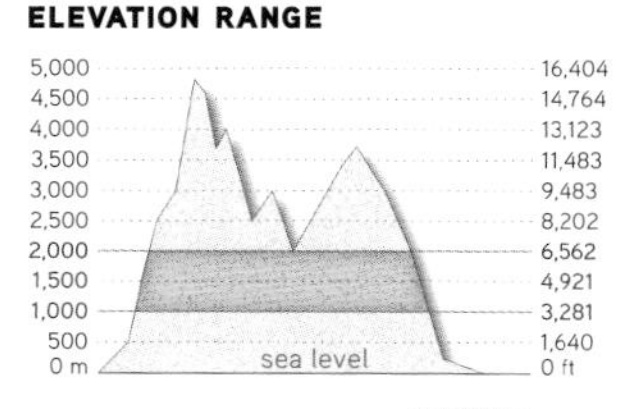

CONSERVATION STATUS LC

VOICE Variable whistled *kwoi-kwoi-eeng* or *kwoi-kweer* or a loud frantic quavered *kwa-a-a-a-ng*

Adult male. Site 26, 23 Jun 07

Leaf holding. Site 26, 23 Nov 08

Female plumage. Site 26, 8 Nov 08

Transitional male. Site 26, 24 Jun 07

SEE ALSO P 104

BRONZE PAROTIA ~ *Parotia berlepschi*

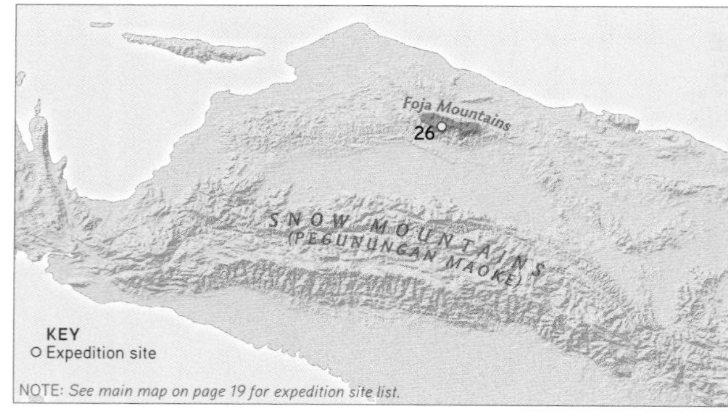

The Bronze Parotia has an incredible story behind its discovery. It resolved a great ornithological mystery surrounding a "lost" bird-of-paradise, spurring a search that spanned more than a century and involved several of the most challenging biological expeditions in New Guinea over the last three decades.

Described in 1897 from two poorly preserved "trade skins" (the skins from dead animals) in the private collection of Count Hans von Berlepsch in 1895, the species went unnoticed at first due to its similarity to Carola's Parotia. But in 1908, three more trade skins surfaced in the collection of Lord Walter Rothschild, and they, too, showed the same darker facial markings and bronzing on the neck and back. The case for calling it a distinct species was strengthened. Unfortunately, nobody knew where the trade skins came from or where this species lived.

Speculation abounded for another 71 years until American ornithologist Jared Diamond conducted a brief helicopter expedition to the higher regions of the uninhabited Foja Mountains of northwestern New Guinea. Diamond saw several female plumaged birds-of-paradise that he identified as Carola's Parotia or a close relative. But without seeing a male, he could only speculate that what he had seen was the lost *Parotia berlepschi*. In 2005, an expedition team led by Bruce Beehler again arrived by helicopter deep in the Foja Mountains. Within short order, the team made a great discovery. They spotted a male parotia similar to Carola's Parotia, but clearly different. It had the dark facial marking and bronze back of the enigmatic trade skins, but it also had pale blue eyes (Carola's Parotias have bright yellow eyes) and its voice was quite distinctive. Eureka! A live *P. berlepschi* had finally been found after 110 years!

In 2007 and 2008, we also had the good fortune to visit the Foja Mountains during a follow-up biological survey to the 2005 expedition. We added more knowledge of this species in the wild through our observations, photos of different plumages, first video of a male at a display court, details of several courtship displays, and additional audio recordings.

ELEVATION RANGE

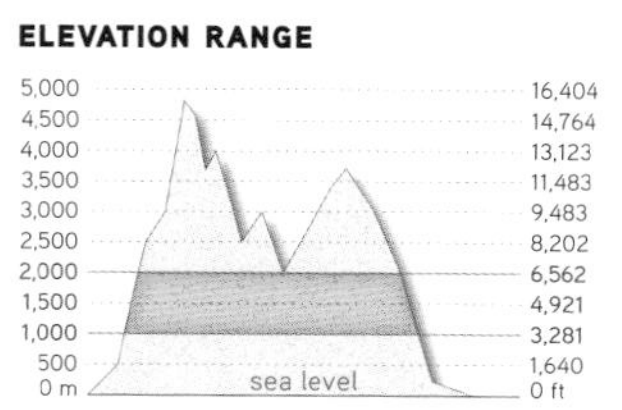

CONSERVATION STATUS LC

VOICE A shrill two-note *wee-deet* reminiscent of a squeak toy

Female and male. Site 7, 1 Dec 04

Ballerina dance. Site 7, 6 Dec 04

Ballerina dance. Site 7, 30 Nov 04

SEE ALSO PP 138, 142-143, 182, 184-185, 186

WESTERN PAROTIA ~ *Parotia sefilata*

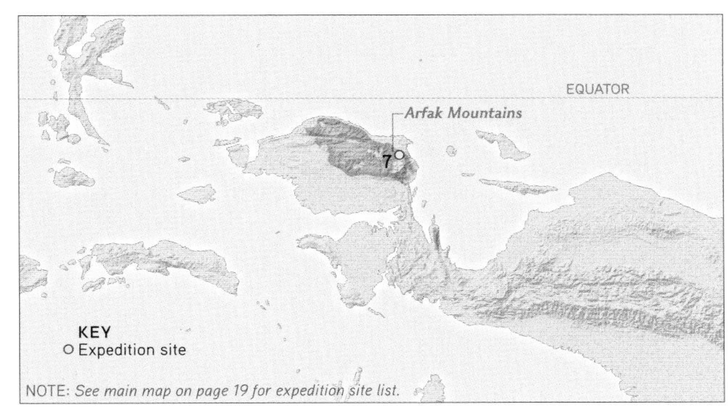

Found only in a few mountain ranges of northwestern New Guinea, the Western Parotia first appeared in an illustration of a poor-quality trade skin in 1781. The species remained unseen for another 90 years until the flamboyant Italian naturalist and explorer Luigi D'Albertis observed (and collected) several individuals in the Arfak Mountains in September 1872. D'Albertis watched, and later recounted in his epic travelogue, a male performing a peculiar dance on a noticeably clear spot on the forest floor: "He spread and contracted the long feathers on his sides, in a way that made him appear now larger and again smaller than his real size, and jumping first on one side and then on the other." Interestingly, D'Albertis interpreted the strange behavior as "an attitude of combat, as though he imagined himself fighting with an invisible foe." We now know that what D'Albertis saw was a practice courtship display, the ballerina dance, being performed on the male's terrestrial display court. Nevertheless, his view underscores just how little the outside world has known about the behavior of most species for nearly 500 years after birds-of-paradise were first discovered.

Our encounters with this species came just over 130 years after those of D'Albertis, not far from where he first explored. One particular court we observed in December 2004 was virtually unchanged nearly five years later, when we returned in August 2009. Sadly, our local informant, Zeth Wonggor, told us that the male was killed by hunters in late 2010. We don't know how long individual display courts are maintained, but this particular one was used for at least six consecutive years, which is the longest documented example of continuous court use.

ELEVATION RANGE

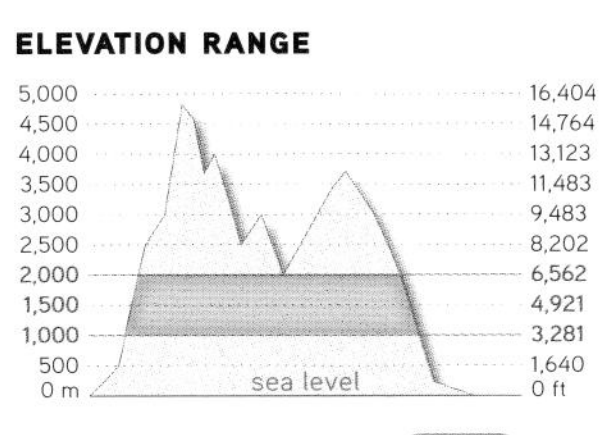

CONSERVATION STATUS LC

VOICE A harsh parrot-like *yaaat,* usually repeated two or three times in succession

Male and female. Site 51, 16 Oct 11

Subadult male. Site 27, 31 Jul 07

Male Display. Site 51, 17 Oct 11

Adult male. Site 27, 31 Jul 07

SEE ALSO PP 105, 124-125, 126-127, 173, 192, 193

WAHNES'S PAROTIA ~ *Parotia wahnesi*

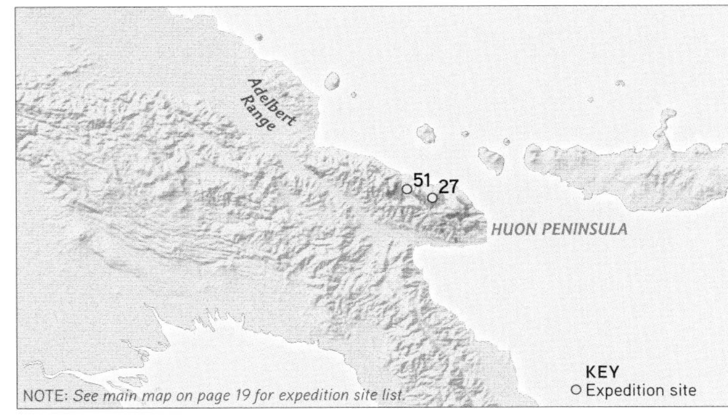

The longest-tailed member of the genus *Parotia,* Wahnes's Parotia is found only in a narrow band of suitable habitat within several mountain ranges of northeastern New Guinea. Its limited distribution occurs in a range of elevation also favored by humans, and for that reason the IUCN Red List considers this species "vulnerable." The isolated population in the Adelbert Range wasn't discovered until 1974 and hasn't been confirmed there since (despite a concerted effort in 2001). If it still exists, its numbers are likely to be extremely small, because the suitable habitat is so limited. The courtship repertoire of Wahnes's Parotia is less complex than that of the other parotias, with only three primary displays; Carola's Parotia, in contrast, has six.

We successfully documented this species on two expeditions, once in 2007 and again in 2011. During the later effort, we worked inside a vast, largely uninhabited forested area called the YUS Conservation Area, which is one of the few successful reserves in Papua New Guinea. This 76,000-hectare (120-square-mile) area was set up to protect the cloud forest home of the endemic Huon Tree Kangaroo, but has the added benefit of protecting three species of birds-of-paradise found nowhere else—including Wahnes's Parotia.

Among the highlights of our latest effort was our successful attempt to photograph and video-record this species from above—that is, from the females' perspective (refer to "Seeing the Female's Perspective" on pages 126-129). This effort to see a parotia ballerina dance led to the discovery of how a small, and usually obscured, patch of iridescent feathers on the top of the head (the "nuchal bar") is used during the waggle phase of the ballerina dance. Once seen from the perspective of the female, we realized that this usually hidden ornament is used to highlight the vigorous back-and-forth waggle movements of the male's head. *Why* this matters is still a mystery, but that it *does* matter to the females is now quite clear.

ELEVATION RANGE

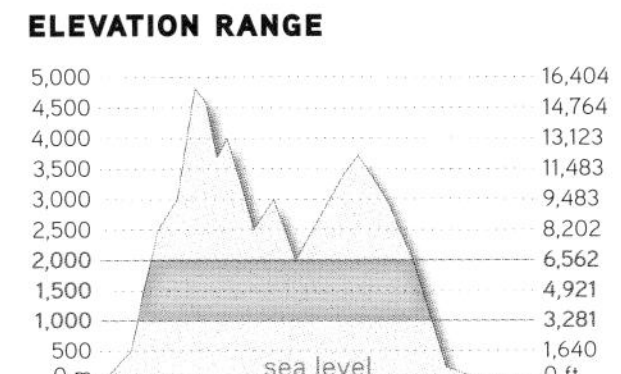

VOICE A harsh, raspy *yeah-yeah* or *yack-yack* similar to a parrot, and a diversity of quieter squeaks and whines given near display courts

Adult male in feeding tree. Site 15, 21 Sep 06

Female-plumaged bird. Site 15, 14 Sep 06

LAWES'S PAROTIA ~ *Parotia lawesii*

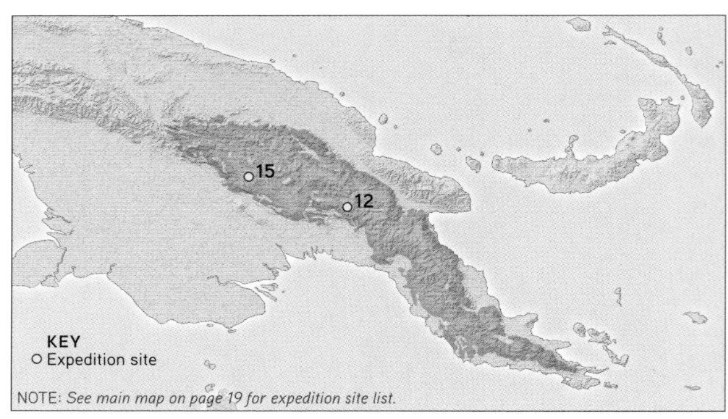

Lawes's Parotia is one of the better studied birds-of-paradise; perhaps only the Raggiana Bird-of-Paradise is more known overall. Yet a well-known bird-of-paradise is a relative term, and many new discoveries are still waiting to be made. For instance, one of the most fascinating displays of this species wasn't revealed to science until 2008. This behavior, which is unique to Lawes's Parotia, is an ensemble version of the more widespread (among *Parotia* species) hops-across-court display.

This species tends to have terrestrial display courts that are tightly clustered, often contiguous, along crests of broad ridges. This means that individual males are forced to interact more than birds with more widely dispersed courts. What evolved from this constant contact is a cooperative—or perhaps merely coordinated—display. In the "group-hop" version of the hops-across-court display, multiple males (up to six have been documented) gather on one court, a practice that is usually taboo and greeted with hostility. The males slowly gather at the periphery of the court and start hopping, in unison, across the length of court, first in one direction and then back again *(view video 45942 at macaulaylibrary.org)*. This dance can go on for several minutes, and all of them head back and forth with a noticeably rigid posture, as if they aren't quite comfortable with one another. Over time, and rather comically, the individual males can get out of sync, until some are hopping in one direction, others are hopping in the other. This causes a "traffic jam" of sorts as they both try to maintain their pace and distance, but inevitably end up bumping into each other and becoming progressively more misaligned as they try to avert collisions. When this happens, the mutual display usually degrades into a flurry of black feathers fleeing the scene and loud angry-sounding calls. Who says scientific discoveries can't be fun?

It should be noted that the easternmost subspecies *(P. lawesii helenae)* has a distinct frontal crest that is smaller and bronze colored. Pending further research, this form may well prove to be a separate species.

ELEVATION RANGE

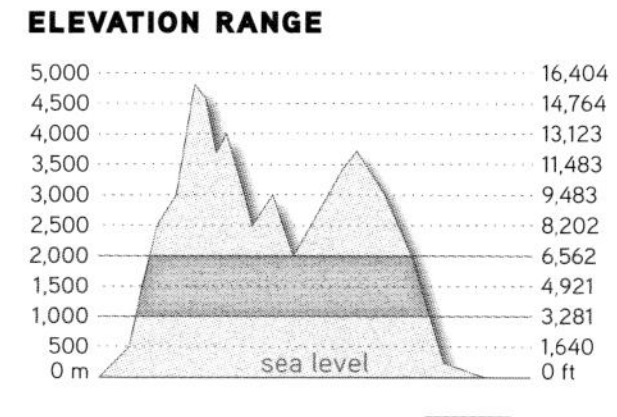

VOICE A loud *nyaack* or *yaack* squawk that is irregularly repeated and not part of a distinct multinote phrase as in other *Parotia* species

---

Male. Site 8, 11 Aug 05

Male. Site 39, 24 Jun 10

Displaying male. Site 39, 16 Jun 10

Female plumage. Site 39, 24 Jun 10

TWELVE-WIRED BIRD-OF-PARADISE ~ *Seleucidis melanoleucus*

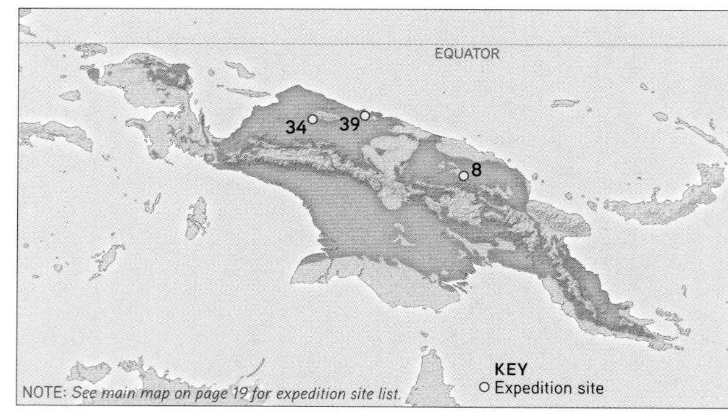

All male birds-of-paradise are genetically "wired" for courtship, but in the Twelve-wired Bird-of-Paradise, this concept has been taken to an unprecedented extreme. Perhaps rivaled only by the outlandish head feathers of the King of Saxony Bird-of-Paradise, the 12 twisted wirelike feathers that protrude from the rear end of a male Twelve-wired Bird-of-Paradise are one of the greatest evolutionary spectacles of the natural world, especially when swiped back and forth across the face of the female during the courtship display! But what's also shocking about these feathers to most people is that the "wires" are not part of the tail. Rather, they are the long bare ends of the bright yellow flank-plumes, which extend from the lower breast and sides of the upper body. The actual tail feathers are merely functional feathers usually tucked away, hidden from view, within the ostentatious mass of wire-tipped yellow plumes.

But the feather curiosities of this species don't end with the wires. The astute student of Latin undoubtedly noticed that the scientific species name for this bird, *melanoleucus*, means "black-and-white"—which is an odd way to name an intensely black-and-yellow bird. The solution to this riddle is a combination of science history and feather biology. The extraordinary yellow color of this species is the result of ephemeral color pigments obtained through eating certain types of fruit that contain the building blocks needed to produce the extraordinary color. But the pigments are inherently unstable and disappear with time if not replenished through the diet. Thus, both living captive birds and long-dead birds-turned-specimens often have completely white flank plumes! The first specimens to arrive in 18th-century Europe were likely trade skins that had not been freshly collected and had, over time, lost their original bright yellow coloration. So, when first named, the moniker "black-and-white" made perfect sense.

ELEVATION RANGE

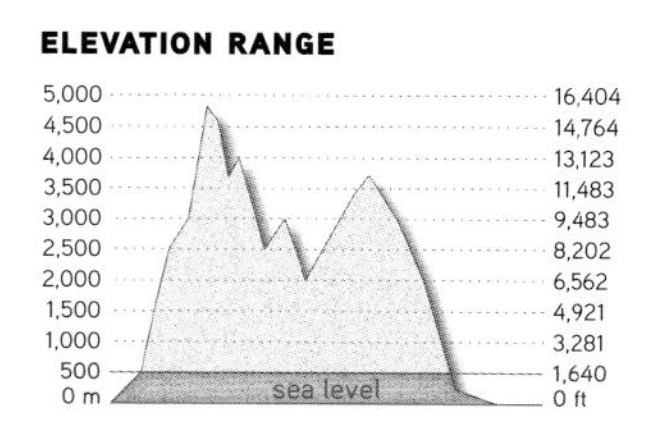

CONSERVATION STATUS LC

VOICE A series of three to eight *waah* or *haw* notes repeated but widely spaced

SEE ALSO PP 6-7, 57, 123, 167, 173

Male performing practice display. Site 11, 30 Sep 05

Male calling (E. Scholes video). Site 11, 29 Sep 05

SEE ALSO P 147

BLACK-BILLED SICKLEBILL ~ *Drepanornis albertisi*

The Black-billed Sicklebill, which was formerly (and is sometimes still) called the Buff-tailed Sicklebill, is one of the more peculiar and not quite beautiful birds-of-paradise. As its common name suggests, one of its defining features is a very long and downwardly curved, sickle-shaped bill. However, the species' scientific name honors the Italian naturalist and explorer Count Luigi D'Albertis, who first discovered the species while exploring the Arfak Mountains of northwestern New Guinea in 1872—not far from where he encountered the Western Parotia (see story on page 205).

Although the Black-billed Sicklebill can be a secretive and difficult-to-spot species, it is nevertheless widespread, relatively common, and usually quite vocal. Yet its elusiveness has led to a lack of knowledge about its courtship display behavior. In 2005, we had the opportunity to make what were, as far as we know, the first-ever photographs and high-definition video recordings of this species' displays.

While spending long hours sitting in a blind to document the displays of Carola's Parotia, we serendipitously discovered a sicklebill's display site. In the later afternoons, we could hear a male Black-billed Sicklebill calling repeatedly nearby, and after extended peering into the surrounding dense forest through the thick leafy walls of the blind, we determined that the sicklebill was using the trunk of a small tree not far off the ground. From a new blind with a clearer view, we later documented several practice displays—that is, ones performed without a female present. The male leaned back until he was perpendicular to the trunk and parallel to the ground, while expanding his long flank plumes around his upper body and head to form a flat, disklike structure facing upward toward the canopy (which is presumably where the female would be watching). Interestingly, on one of the few other occasions in which this species has been seen to display, it was also using the side of a small tree in close proximity to a Carola's Parotia display court. The reason for this association is unknown, but it is unlikely to be a mere coincidence.

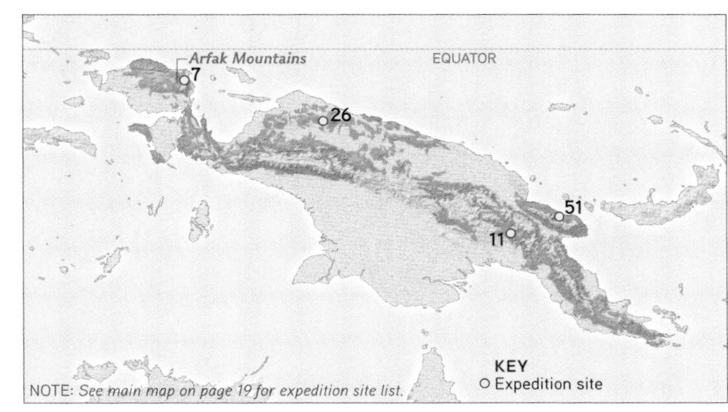

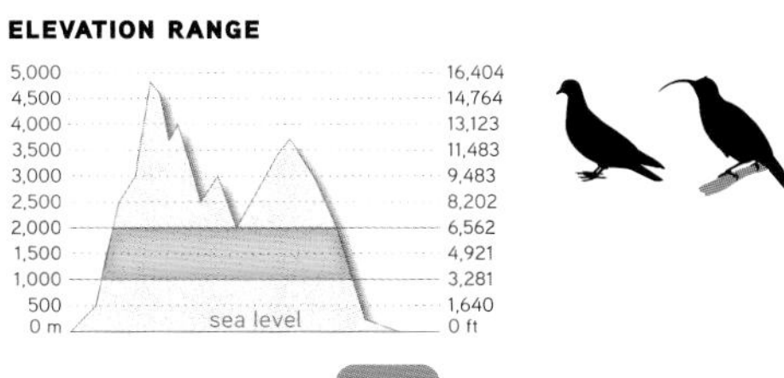

CONSERVATION STATUS LC

VOICE A powerfully whistled and descending *dyu-dyu-dyu-dyu-dyu* that increases in speed as delivered

---

Male feeding. Site 39, 30 Jun 10

Female. Site 39, 29 Jun 10

Male feeding. Site 39, 30 Jun 10

SEE ALSO P 55

PALE-BILLED SICKLEBILL ~ *Drepanornis bruijnii*

The Pale-billed Sicklebill might even be more peculiar and less beautiful than its sister species the Black-billed Sicklebill. It shares a similarly decurved, sickle-shaped bill, but in this species it is even more pronounced because of its ivory white coloration. The Pale-billed Sicklebill is also one of the least ornamented of the sexually dimorphic "plumed" (i.e., non-manucode) birds-of-paradise. Unlike the Black-billed Sicklebill, the flank plumes of the male are small and not known to fan out prominently, although many details of their courtship displays are still lacking. Its most conspicuous ornaments are the elongated and iridescent-tipped breast feathers (pectoral fans) and the unusual feathering of the head.

More extensively than its sister species, the Pale-billed Sicklebill has a defined region of bare gray-purple facial skin around the eyes, which is a unique type of nonfeather ornament. When viewed head-on and from above, these bare patches give the male the appearance of having a Mohawk or wide "racing stripe" along the top of the head.

Another hallmark of this species is how difficult it is to see—it is arguably the most cryptic and elusive of all the birds-of-paradise. But like its secretive sister species, the Pale-billed Sicklebill is quite vocal and usually located only after first hearing its distinctive sound. Its voice is quite unlike that of any other bird-of-paradise in that it's a rather long series of odd and irregularly phrased whistled notes that ramble on well beyond expectation.

The Pale-billed Sicklebill is also much more narrowly distributed than the Black-billed Sicklebill. It is found only in a relatively small area of lowland forest and swamp forest in the northwestern corner of New Guinea. While we didn't document this species' courtship displays, Tim did manage to get unique photographs of both males and females using their exceptional bills while foraging for fruit in the canopy (see "View From the Canopy" on page 98).

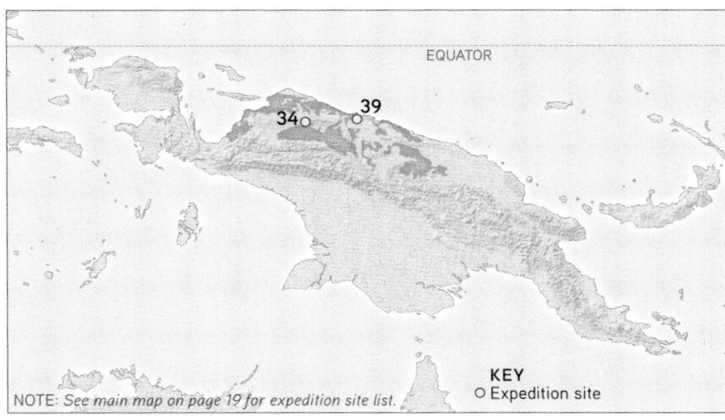

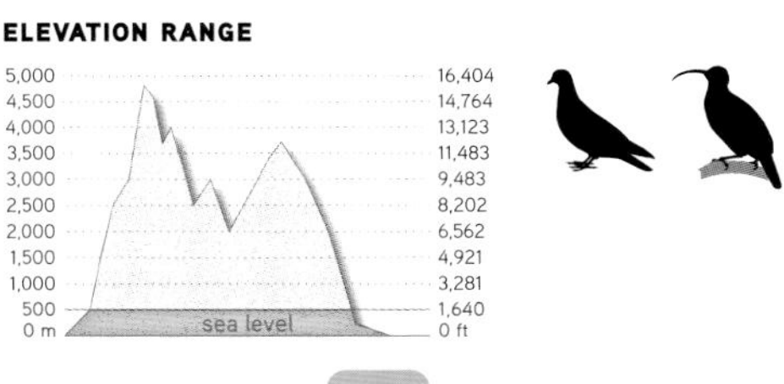

CONSERVATION STATUS NT

VOICE Loud, hollow, whistled notes with a musical quality, but very irregular in delivery

Male aerial display dive. Site 30, 23 Jul 08

Male. Site 28, 15 Jul 08

Female. Site 30, 23 Jul 08

SEE ALSO PP 60, 85, 146, 173

STANDARDWING BIRD-OF-PARADISE ~ *Semioptera wallacii*

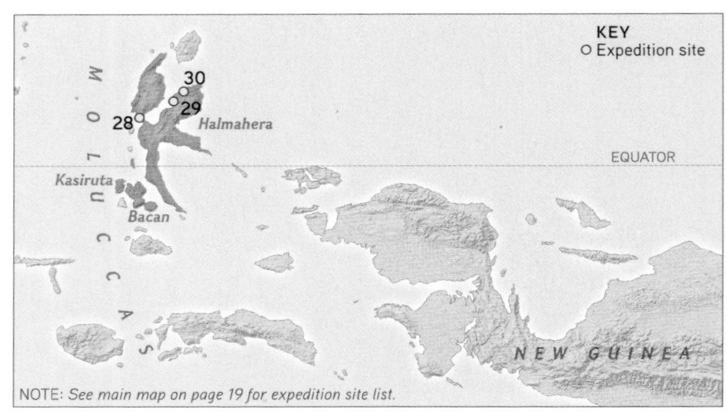

The Standardwing Bird-of-Paradise is found only on the islands of Halmahera, Bacan, and Kasiruta in the northern Moluccas. As one of only two birds-of-paradise to inhabit the Moluccas (the other being the Paradise-crow), it has the westernmost distribution of any plumed bird-of-paradise.

Alfred Russel Wallace first discovered it during his visit to the spice island of Bacan in October 1858. Soon after arriving, Wallace sent his assistants out to shoot birds while he went to explore the nearby forests. When he returned, an assistant presented a specimen that astounded Wallace. It was, as Wallace writes in his epic travelogue, "a bird with a mass of splendid green feathers on its breast, elongated into two glittering tufts; but, what I could not understand was a pair of long white feathers, which stuck straight out from each shoulder." Wallace was ecstatic as he came to realize the significance of his discovery. "I now saw that I had got a great prize, no less than a completely new form of the bird-of-paradise, differing most remarkably from every other known bird."

The Standardwing Bird-of-Paradise is indeed unusual. For starters, the main male ornament, the wing "standards" (meaning something supported in an upright position, such as a flag) are truly exceptional. They are two greatly elongated feathers emerging from each wrist that can be lifted straight up and waved over the back during courtship—including during a very bizarre flight display.

In 2008, we had the good fortune to document a courtship display area (also known as a lek) for this species on Halmahera. Local guides helped us build a canopy platform tied between two trees. From there, we observed the communal displays of several males in near darkness as the sun started to rise. During these displays, we saw the strange "parachute" display flight, in which a male suddenly, amid the cacophony of the other displaying males nearby, launches himself skyward from a branch until he emerges above the treetops and then flutters, like a butterfly, back down to the same perch while waving his four white wing-standards like flags over his back.

ELEVATION RANGE

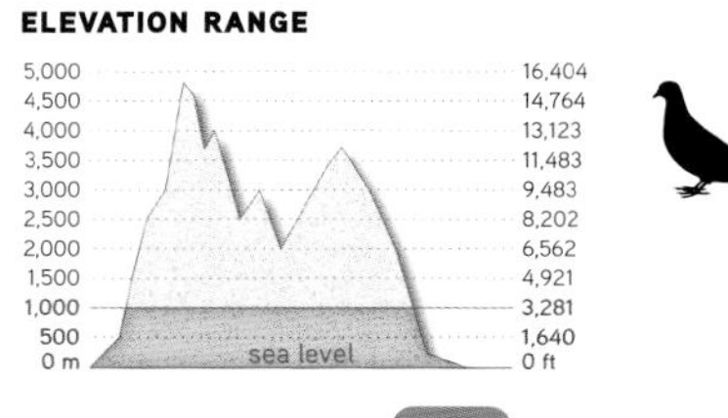

CONSERVATION STATUS LC

VOICE Repeated loud nasal upslurred *wark* or *wau-wau* notes

Female plumage. Site 23, 11 Dec 06

Adult male. Site 15, 30 Nov 11

Female plumage. Site 15, 1 Nov 11

Male display (E. Scholes video)

SEE ALSO PP 170–171, 194, 195

SUPERB BIRD-OF-PARADISE ~ *Lophorina superba*

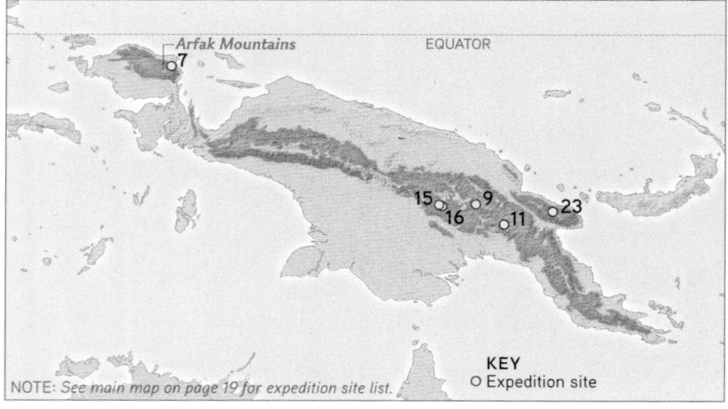

The names of birds-of-paradise are not short on superlatives, but the Superb Bird-of-Paradise truly deserves its moniker. Described in 1781, using a trade skin obtained in the Arfak Mountains of northwestern New Guinea, this distinctive species is the only member of the genus *Lophorina*. Common and widely distributed throughout the mountainous forests of mainland New Guinea, it tolerates human disturbances and is often found in forest patches near villages.

The most conspicuous features of the male Superb Bird-of-Paradise are the bizarre "cape" of long velvet-black feathers that extend from the back of the head, the iridescent blue crown patch on the top of the head, and the striking delta-shaped iridescent metallic-blue breast shield. But this species also has another, more subtle ornament in the form of two unusual tufts of elongated movable feathers at the base of the bill. These bill tufts play a central role in the extraordinary transformation that takes place during courtship.

The bizarre act of shape-shifting for which this species is renowned is unrivaled even among the other body-form-altering birds-of-paradise. Interpretations among those who've seen the display for the first time vary from "space alien" to "psychedelic smiley face" (a personal favorite). The hard-to-describe form is produced by an almost magical interaction of the three prominent ornaments—cape, crown, and breast shield—and the special bill tufts. The smiley-face form emerges when the head is tilted back, bill pointed skyward, and the cape is spread flat across the back and pushed forward over the head. As the cape goes forward, the blue crown-patch also gets pushed forward to become transposed against the black backdrop of the now forward-facing, conical-shaped cape. Viewed from the front, the crown is bisected by the upward-pointing bill and appears as two bright (nearly white) circular eyes. Thus, the round eyes of the smiley face are produced by the bill tufts, which curve away from the upward pointing bill to frame the crown patches and create the illusory eyes that seem to magically appear from out of nowhere above the blue smile of the breast shield. It's a superb trick indeed!

ELEVATION RANGE

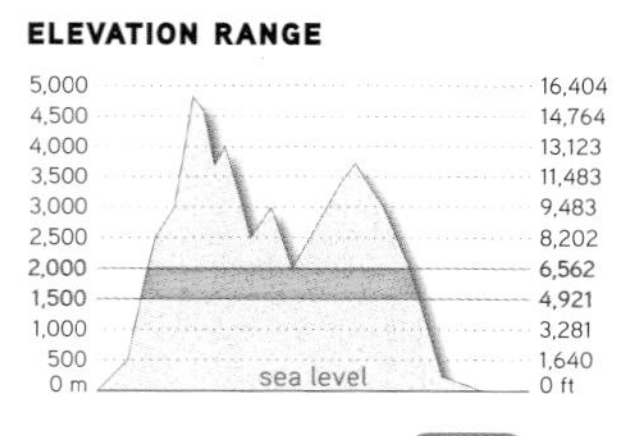

CONSERVATION STATUS LC

VOICE Loud series of shrill *shraak* or *scree* notes that are variable in number and sometimes drawn out in longer *shreeeeee* notes

Young males practice displaying. Site 46, 23 Nov 10

Female plumage. Site 46, 20 Nov 10

Male with fruit. Site 46, 19 Nov 10

SEE ALSO PP 46, 139

PARADISE RIFLEBIRD ~ *Ptiloris paradiseus*

The more southern of the two Australian riflebirds, this species is also the southernmost bird-of-paradise. Though it is found in the subtropical and temperate rain forests of eastern Australia, its current distribution is mostly limited to hill forest, because much of its original habitat has been cleared for human development. Human population growth, combined with the large numbers of birds killed during the hunting at the turn of the 20th century for hat plumes, led some experts to believe that the species was faced with the imminent threat of extinction. Yet the Paradise Riflebird has managed to hang on. Today it can be seen with relative ease in parks and recreational areas on the outskirts of major urban areas.

We documented this species in a patch of regenerated forest on a broad hilltop that provided a distant view of the city of Brisbane. Fortunately, the area is now surrounded by a national park, and most of the landowners have traded farming for ecotourism and weekend getaway cottages. But given its close proximity to a major city and relative ease of access via well-maintained roads, campgrounds, and information from the local birding community, this species is one of the easiest birds-of-paradise to experience in the wild. (Its sister species Victoria's Riflebird, found a bit farther north, also is easy to find.)

For that reason, it is surprising that details of its courtship display are not that well known. The ritual is very similar to its better known sister species to the north, but the ways in which they differ are not entirely clear. It seems this species prefers wide horizontal limbs for courtship displays whereas Victoria's Riflebird prefers the top of a tall, broken tree stump or pole—although the Paradise Riflebird is also known to use a pole on occasion. The main display involves the male standing upright and raising his unusually shaped wings, which are ornamentally rounded with blunt-tipped flight feathers, over his head while performing a rhythmic "fan dance" in which his head is alternately concealed behind each wing as he rapidly moves it from side to side in coordination with the raised wings.

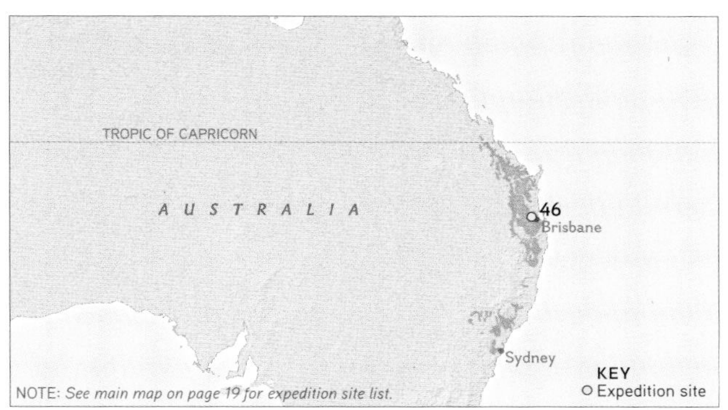

ELEVATION RANGE

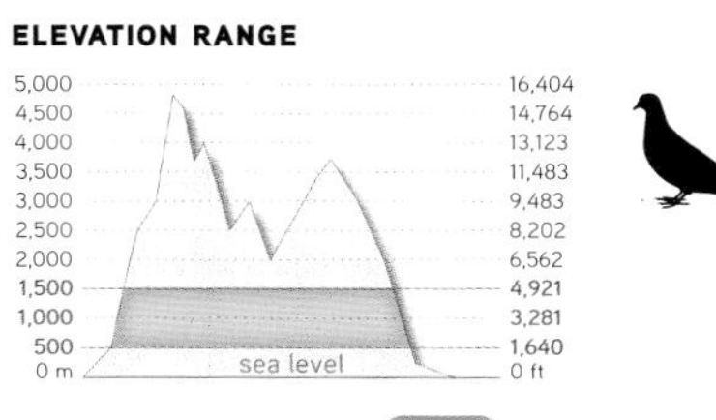

CONSERVATION STATUS

VOICE Raspy, explosive *yaasss* note repeated twice in succession

Calling male. Site 31, 5 Sep 08

Male. Site 31, 4 Sep 08

Display to female. Site 31, 5 Sep 08

Female. Site 31, 3 Sep 08

SEE ALSO PP 47, 134, 154

VICTORIA'S RIFLEBIRD ~ *Ptiloris victoriae*

The smallest of the three riflebirds, Victoria's Riflebird is found only in the Atherton region of tropical northern Queensland, Australia. It was discovered by John MacGillivray, son of ornithologist William MacGillivray, in 1848 while serving as the naturalist aboard H.M.S. *Rattlesnake*, which explored the Australian coast under the command of Captain Owen Stanley. Famed ornithologist John Gould named the species in honor of Queen Victoria in 1850. Because its distribution lies in one of the most accessible and visited regions of northern Australia, Victoria's Riflebird is fairly well known. Sometimes it can be found around the gardens of private homes and eco-lodges bordering the rain forest.

Its primary courtship behavior consists of a circular-wings display in which the male faces the female on his tree-stump display pole with his rounded wings open and raised to meet above his head. From this pose, the male repeatedly raises and lowers himself with his legs in a slow, rhythmic way, all the while tracking the female if she moves left or right. Although no vocal sounds are given, the male stands with his mouth wide open in a "gaping display" to show off its bright yellow interior.

If the female arrives on the top of the display pole, he leans to one side and then the other while doing a "fan dance" with his wings to alternately hide his head behind each wing. The pace of the fan dance increases in intensity the longer the female stays to observe the display. By the end, the male is rocking vigorously back and forth, leaning toward the female until he is literally clapping her between his wings with each head-to-wing motion. If the female is ready to mate, she will remain on the display pole, and the male will suddenly stop clapping her with his wings and mate with her. After mating, or especially if the female flies off before mating, the male will often adopt the open-wing upright posture and hold it for an extended period before slowly letting his wings drop to his sides. The entire display is quite dramatic and a wonderful spectacle to behold. And given its relative accessibility in a well-traveled part of Australia, it comes highly recommended!

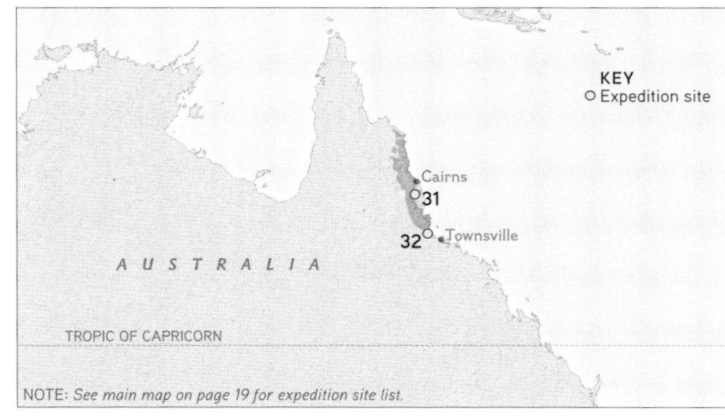

ELEVATION RANGE

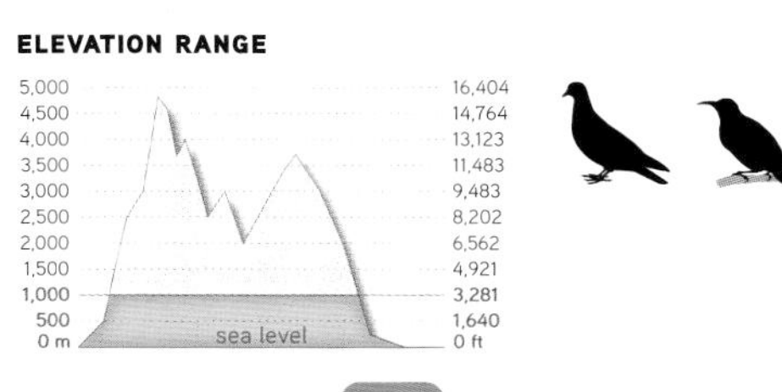

CONSERVATION STATUS LC

VOICE An explosively loud and harsh-sounding *yaasss* note like the Paradise Riflebird's but not always delivered twice in succession

Male calling from display vine. Site 38, 24 Aug 09

Display to female. Site 38, 21 Aug 09

Male. Site 38, 22 Aug 09

MAGNIFICENT RIFLEBIRD ~ *Ptiloris magnificus*

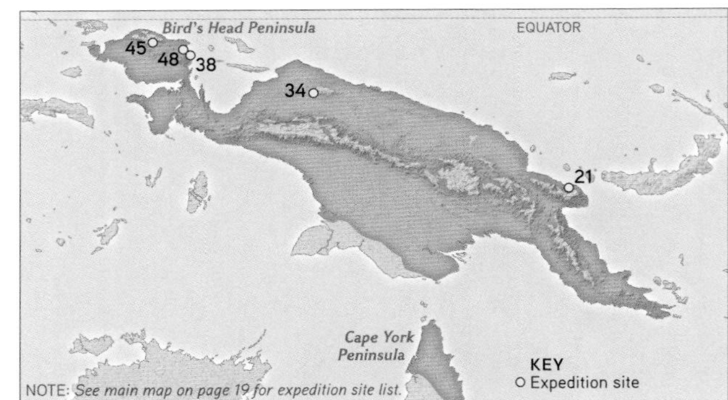

One of the most widespread birds-of-paradise, with a distribution approaching that of the Trumpet Manucode (which is *the* most widespread), the Magnificent Riflebird is one of only two species of birds-of-paradise found in both New Guinea and Australia. In the former, it is found in lowlands and hill forests virtually throughout the island. In Australia, it occurs in the rain forests of the Cape York Peninsula in the far northeast. While similar in appearance to the two endemic Australian riflebirds, the Magnificent Riflebird is considerably larger with a bright metallic blue iridescent breast instead of the iridescent greenish version found in the Australian-only species. Like all the riflebirds, males have wings of an unusual rounded shape that have evolved under the influence of sexual selection to serve as ornaments during courtship in addition to their more normal use for flight.

While we encountered this species at several field sites, we focused on it in the lowland forests of northeastern New Guinea's Bird's Head. Peninsula. It was quite common and very vocal, with its clear, powerful whistle resonating throughout the forest from dawn to dusk. Shortly after setting up camp, we discovered a thick, nearly horizontal vine hanging just above arm's reach in the understory that the nearby male used for courtship display. After building a well-camouflaged blind on the ground nearby, we waited to see and document the extraordinary display. The male rapidly snaps his head from side to side, from behind one outstretched wing to another, while bouncing up and down and moving back and forth along the top of the vine. Our patience was only rarely rewarded, however. While we saw the bird perched and calling from his vine on a daily basis, sometimes for hours, we nevertheless saw only three displays in over ten days of sitting and waiting between the two of us!

It is also worth noting that the easternmost New Guinean subspecies (*P. magnificus intercedens*) has a distantly different vocalization—more of a growl than a whistle—and upon further investigation might well be considered a separate species.

ELEVATION RANGE

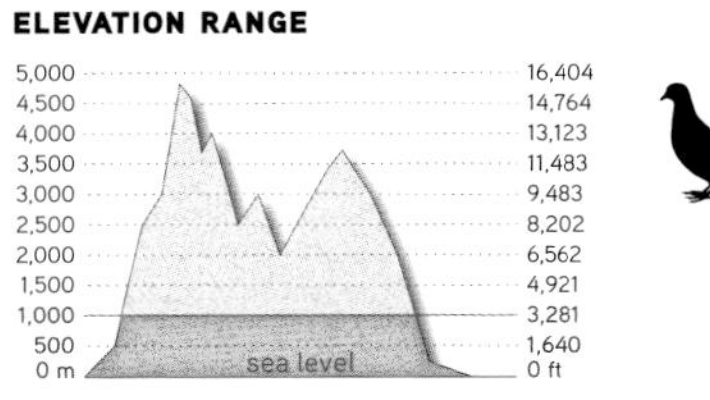

CONSERVATION STATUS

VOICE Geographically variable—either a powerful, clear, upslurred *woiieet-woot-woi-ieet* whistle or a growling *eerrau-eerrau-wow*

SEE ALSO PP 71, 119, 188, 189

Male predisplay. Site 26, 25 Jun 07

Female plumage. Site 15, 28 Oct 11

Male display. Site 26, 25 Jun 07

BLACK SICKLEBILL ~ *Epimachus fastosus*

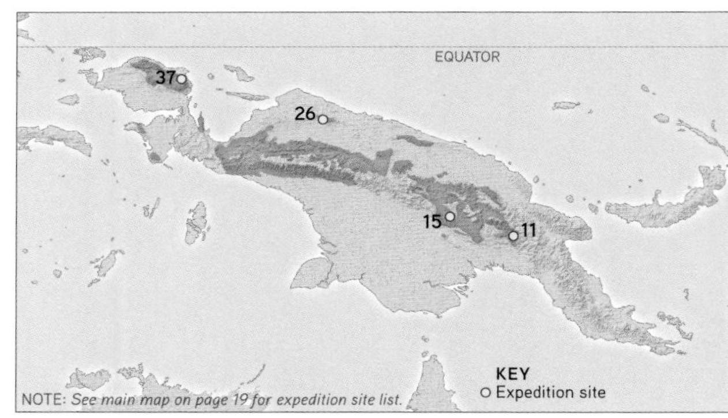

With its long sickle-shaped bill, long tail, and mostly black plumage, the Black Sicklebill proved to be an enigma for early naturalists. For decades, its distinctive bill led ornithologists to think it was related to birds with similarly shaped bills, such as the African woodhoopoes (Phoeniculidae). Not until the late 19th century did scientists determine that it was actually a bird-of-paradise.

Found in the mountains of western and central New Guinea, the Black Sicklebill is one of the largest birds-of-paradise. Because it lives in forests within a zone of elevation also favored by humans, its saber-shaped tail feathers are regularly featured as centerpieces on the headdresses of many traditional highland cultures. But the real threat to the species is habitat loss, not overhunting.

Its unusual and far-carrying voice is a quintessential cry of mountain forests within its range. Yet for most of the last 200-plus years that the species has been known to science, very little was known about its courtship behavior. Early scientific illustrators had difficulty imagining how the bizarre axe-shaped "epaulette-like" feathers of the upper breast were used. Most depictions show them outstretched, winglike, to the sides while the bird displays in a horizontal posture. But like many birds-of-paradise, reality is even stranger than the imagination.

When displaying, males perch upright on a broken tree stump, or pole, in the forest and fan the epaulette-like plumes up and out so that they entirely surround the head, turning the upper body into a black ovoid shape. As each individual feather of the "fan" falls into place, an extraordinary meta-ornament appears in the form of a brilliant metallic-blue line framing the ovoid. When the female approaches the display pole, the male leans entirely to one side and slowly moves his now horizontal ovoid body up and down with his legs. As the female lands next to him, the male snaps upright and looms over her like a phantom, rocking back and forth and side to side as she scrutinizes him from close range. The overall effect is striking and otherworldly.

ELEVATION RANGE

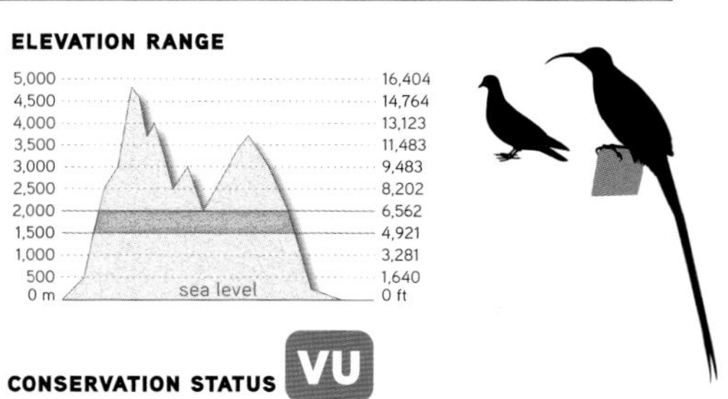

CONSERVATION STATUS VU

VOICE A very loud and fluid-sounding two-note *quink-quink* or a faster four-note *wik-wik, wik-wik*

SEE ALSO PP 190, 196, 197

Male display. Site 2, 5 Sep 04

Female feeding. Site 1, 8 Oct 05

Adult male. Site 18, 26 Oct 11

SEE ALSO P 33

BROWN SICKLEBILL ~ *Epimachus meyeri*

Although similar to its sister species the Black Sicklebill, the Brown Sicklebill is a little smaller, has a pale eye, and features more brown on the breast and belly. Vocally, though, the two are easily to tell apart: While both have loud calls that are characteristic sounds of the forests they inhabit, the male Brown Sicklebill makes one of the most remarkable vocalizations in all of the natural world. It doesn't sound like a bird—or even an *animal*, for that matter. Instead, the sound is best compared to the burst of a machine gun or the pounding of a jackhammer—at a volume nearly as loud as either machine. The rapid-fire vocal burst is so like a machine gun that during World War II, when Japanese soldiers were first making their way into the vast mountains of the interior, the sound of this bird-of-paradise reportedly fooled them into thinking they were coming under fire in an ambush. Having heard this species numerous times ourselves at both close range and from afar, we can well believe the story.

Throughout much of its range, the Brown Sicklebill overlaps with the Black. In areas where both species occur, they segregate by elevation, with the Brown found at the higher elevations. But little is known about the factors that keep the two species separate in areas of overlap.

Courtship displays of this widespread species were unrecorded in the wild when we began our project; they were only partially known from captive birds. We had the chance to see several displays near Mount Hagen (the mountain, not the nearby town) in Papua New Guinea. One big question was whether this species leaned fully horizontal when in the ovoid display, as does the Black Sicklebill. As we discovered, they do indeed lean fully horizontal, even though they display from a horizontal branch instead of a pole like the Black.

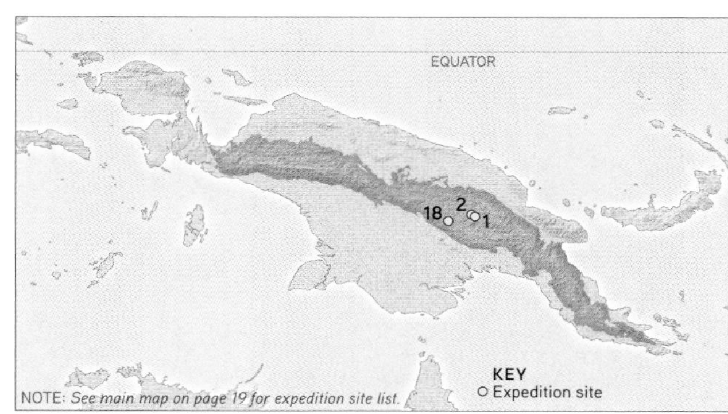

ELEVATION RANGE

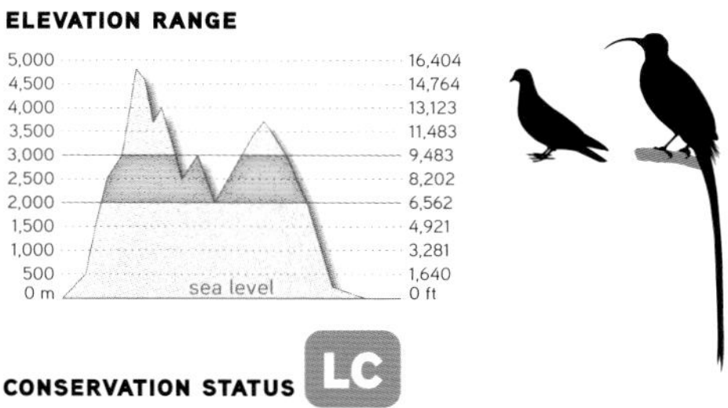

CONSERVATION STATUS LC

VOICE An unbirdlike dry, hollow burst, *tat-at-tat-at-at-a-at,* like a machine gun

Adult. Site 7, 4 Jul 10

Sitting on nest. Site 7, 5 Jul 10

Sitting on nest. Site 7, 5 Jul 10

SEE ALSO P 43

LONG-TAILED PARADIGALLA ~ *Paradigalla carunculata*

One of the least well known birds-of-paradise, the Long-tailed Paradigalla was the first species in its genus to be described, in 1835 by French naturalist René Primevère Lesson. The genus name *Paradigalla* loosely means "paradise chicken" (chickens are in the genus *Gallus*) and was given because the most defining features of the two *Paradigalla* species are the prominent fleshy facial wattles, which Lesson apparently likened to the wattles of a chicken.

This species has been found only in the Arfak Mountains of far western New Guinea, but likely also lives in the forests of the other mountain ranges in the Bird's Head Peninsula. Like its sister species the Short-tailed Paradigalla, the Long-tailed Paradigalla is unusual among birds-of-paradise in that the sexes are very similar in appearance. The males possess few ornamental plumes, and the ornamental facial wattles mark both sexes. The most distinguishing feature of this species, as its name suggests, is that it has a much longer tail than its nearest relative, which has an unusually short tail.

Very little is known about the diet and foraging behavior, voice, courtship behavior, and breeding biology of this species, but presumably all are the same or quite similar to the Short-tailed Paradigalla. Our observations, photographs, and video of this species are the first-ever documentation of its nest. While it was very similar in size, placement, and construction to that of the Short-tailed, we gathered valuable scientific data nonetheless.

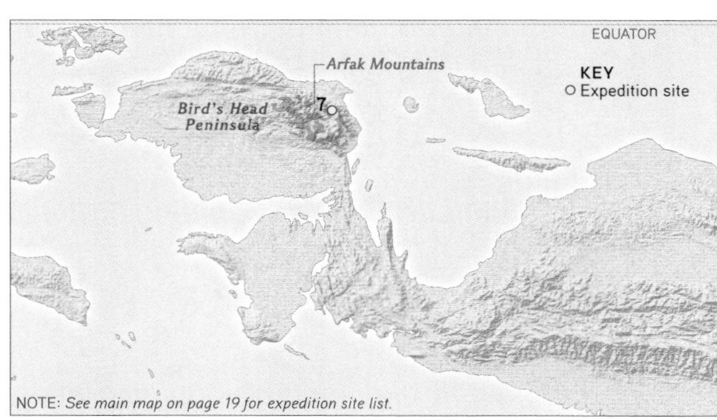

ELEVATION RANGE

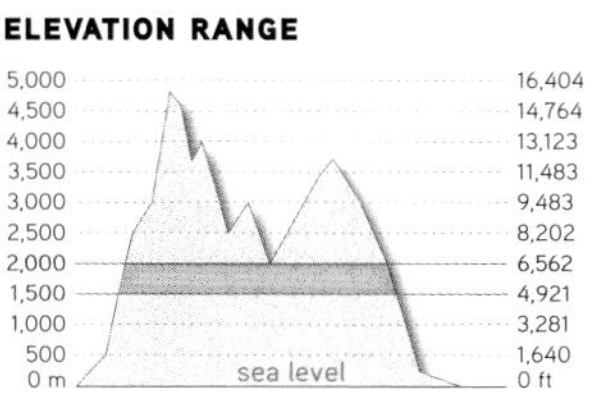

CONSERVATION STATUS

VOICE Little known, but possibly a clear monotone whistled *wheeeeee*

SHORT-TAILED PARADIGALLA ~ *Paradigalla brevicauda*

Adult. Site 16, 30 Nov 10

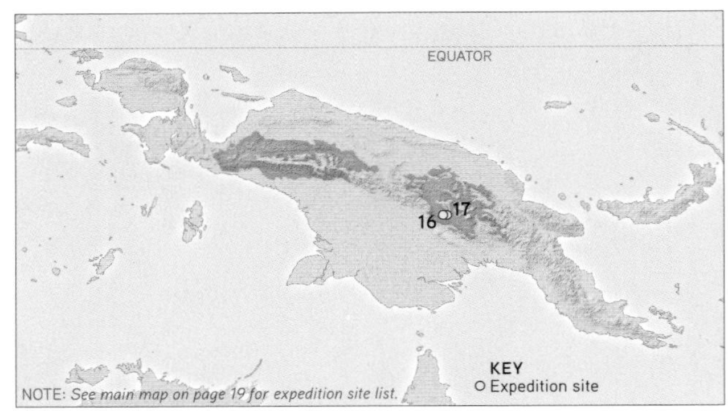

Found in the mountain forests of western and central New Guinea, the Short-tailed Paradigalla, like its sister species the Long-tailed Paradigalla, is an exception among the plumed birds-of-paradise: The males and females are very similar, not possessing much sexual dimorphism (differences). Males are jet black and stocky, with a very short tail and little ornamentation beyond a bit of iridescence on the head and, most prominently, large yellow, small blue, and tiny red wattles on the forehead and face. Females are only slightly duller, but otherwise appear the same.

Because of their lack of sexual dimorphism and ornamental plumes, the two *Paradigalla* species are often regarded as an early fork of the bird-of-paradise family tree, branching off just after the Paradise-crow and manucodes. But scientists have also long believed that it is closely related to the genus *Astrapia*. Recent DNA-based evidence reveals that the genera *Paradigalla* and *Astrapia*, while closely related, are not an early branch off the bird-of-paradise family tree, but instead are a branch firmly within the core group of plumed species. What this implies is even more interesting: Through the course of evolution, the ancestors of the two *Paradigalla* species apparently lost their sexual dimorphism and ornate plumes. But because so little is known about the courtship and mating behaviors of the two species, we don't know if this evolutionary trait occurred because sexual selection has acted in reverse to the way we normally think it does—made being *less showy* the more attractive option—or if a factor in the reproductive biology of bird-of-paradise changed so that the requirements of sexual selection have been scaled back and natural selection has started to bring these species back into the realm of "normal" birds. Hopefully, future research will shed light on this interesting question.

ELEVATION RANGE

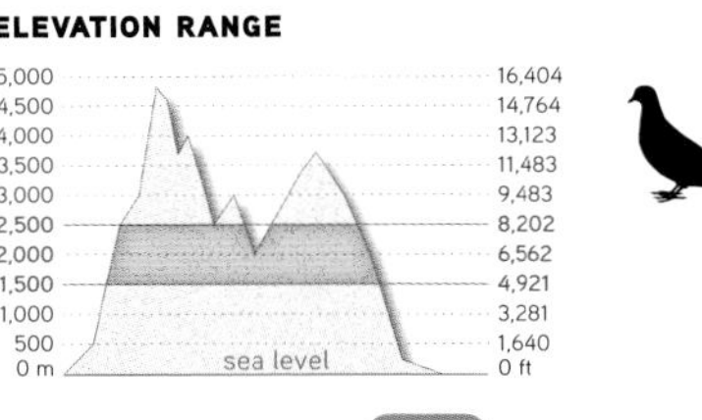

CONSERVATION STATUS

VOICE Little known; some records of a high-pitched, mournful four-note whistle or *hui* note

SPLENDID ASTRAPIA ~ *Astrapia splendidissima*

Adult male feeding on schefflera. Site 40, 20 Jun 10

Adult male. Site 40, 20 June 10

Adult male. Site 40, 20 Jun 10

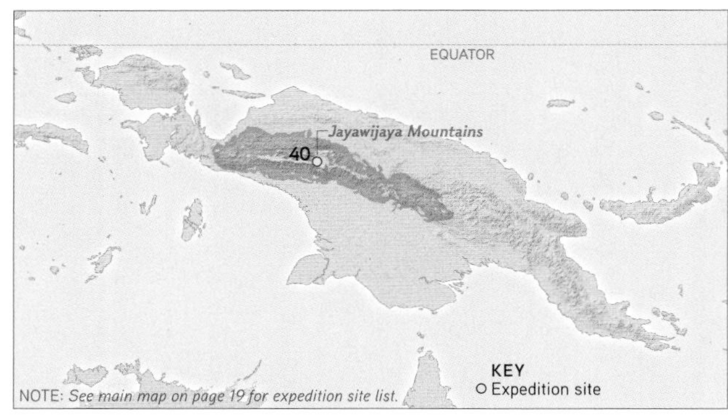

Although it is the shortest-tailed member of the "long-tailed" bird-of-paradise genus *Astrapia*, the Splendid Astrapia nevertheless earns its name from its colorful plumage. The most iridescently feathered bird-of-paradise, it is in fact one of the most iridescently feathered birds in the world (rivaled only by several hummingbirds, sunbirds, and pheasants). As a colleague of ours once put it, the Splendid Astrapia "looks like a disco ball in a tree." While all astrapia males have considerable iridescence, the intensity and coverage of its brilliant metallic blue-green plumage is unparalleled, as is the color and intensity of the metallic orange-red on the neck and throat.

The Splendid Astrapia is distributed relatively widely throughout the entire western half of New Guinea's central cordillera, yet very little is known about it, which is true of most of the *Astrapia* species. It prefers to inhabit the high-elevation mountain forests above 2,000 meters (6,600 feet), where conditions are cold and very wet. We observed this species in edge-forest habitat along the shores of Lake Habbema in the Snow Mountains of Papua, Indonesia, which is around 3,000 meters (10,000 feet) in elevation, and an adult male feeding on schefflera fruit was photographed above 3,300 meters (11,000 feet).

There is also a dearth of information about courtship displays in this species. During our limited period of study, we observed a few behaviors that were probably courtship displays, but they were distant and given without a female present. The display behavior we observed was very similar to one we've observed in the Ribbon-tailed Astrapia, in which the male adopts a "hunchback" horizontal posture on a horizontal branch and pivots from side to side before making rapid hop-flights back and forth between several adjacent perches.

ELEVATION RANGE

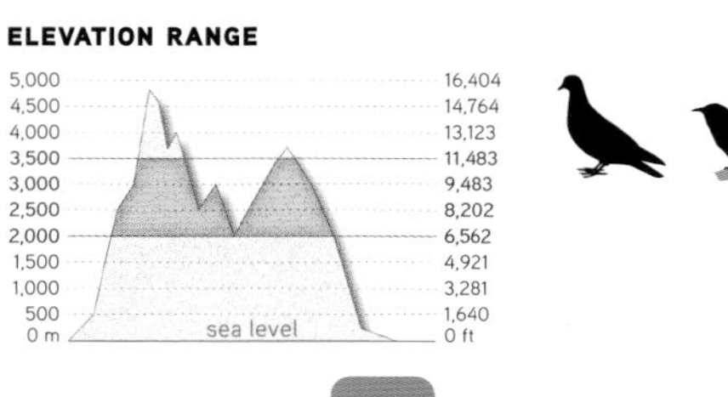

CONSERVATION STATUS LC

VOICE Little known; a *to-to-ki* or a *jeert-jeert-tink*, sounding rather like a frog or insect

SEE ALSO PP 8–9, 169

Subadult male. Site 37, 3 Sep 09

Inverted display. Site 37, 3 Sep 09

Subadult male. Site 37, 3 Sep 09

SEE ALSO PP 152-153

ARFAK ASTRAPIA ~ *Astrapia nigra*

Although first described from a trade skin in 1788, the Arfak Astrapia wasn't found in the wild for nearly another century, when Italian botanist Odoardo Beccari, on an expedition with Luigi D'Albertis, encountered it living in the higher reaches of Arfak Mountains of western New Guinea in 1872. This bird-of-paradise is found only in the higher-elevation forests of the Bird's Head Peninsula. While not uncommon within its range, the species has nevertheless been infrequently spotted, especially its adult males. Among several species that vie for the title, the Arfak Astrapia may be the least known bird-of-paradise.

Until 2009, when we observed and documented several practice displays by a subadult male in transitional plumage, nothing was known about the courtship behavior of this species (see "New Discoveries: *Astrapia*" on pages 150-153). While no displays to females have been observed so far, the general courtship display behavior resembles very much that of the Huon Astrapia. The young male we observed rotated himself below a horizontal perch by dropping, tail first, to hang upside down below the perch. From this inverted posture, the male spread his tail feathers and cocked his tail so that it pointed skyward. Like the Huon Astrapia, the displaying male also lunged his head and bill upward repeatedly while hanging upside down. Although no female was present, the similarity to the display of the Huon Astrapia makes it all but certain that the lunging is directed toward a female that would perch just above the male on the same branch and look down at his flaglike tail and brilliant green undersides (a plumage shared by both the Huon and Splendid Astrapias).

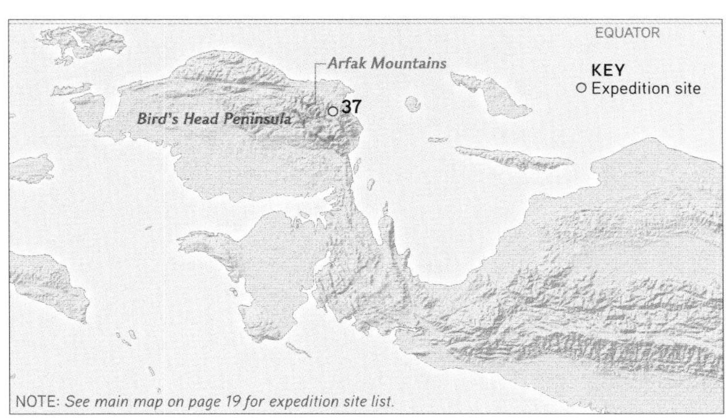

ELEVATION RANGE

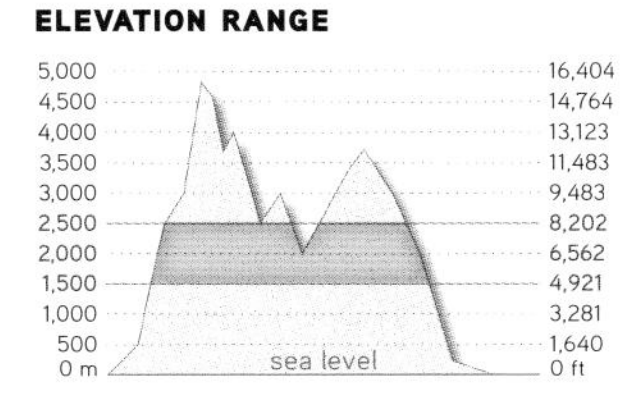

CONSERVATION STATUS LC

VOICE Little known; a dry hollow click or *cluck* sound that is reminiscent of a human tongue click

Female feeding. Site 23, 11 Dec 06

Male feeding. Site 20, 5 Dec 06

SEE ALSO PP 53, 106, 151

HUON ASTRAPIA ~ *Astrapia rothschildi*

Found only in the Huon Peninsula in northeastern Papua New Guinea, the Huon Astrapia is similar to the Arfak Astrapia in overall appearance. It also lives only in one of New Guinea's isolated outlying mountain ranges. In addition, it is another of the more poorly known birds-of-paradise. Prior to our observations and documentation, only one other expedition had observed the courtship display behaviors of this species in the wild, although the New York Zoological Society in the 1930s had made observations of their birds' displays in captivity.

We had the opportunity to document this species in the high-mountain forests of one of Papua New Guinea's most successful conservation efforts, the YUS Conservation Area, which covers 76,000 hectares (120 square miles) of tropical forest from the north coastal lowlands to higher mountains of the interior. The Tree Kangaroo Conservation Project has worked with local landowners for more than a decade in order to protect the native forest of the Huon Tree Kangaroo, which like the Huon Astrapia is found only in the mid- to high-elevation forests of Huon Peninsula.

One of the most interesting aspects of the courtship displays we uncovered is what happens just prior to and following mating. During the premating display, which was unknown, the male probes the female with his bill in a ritualized way while hanging upside down from a horizontal branch. If the female continues to be interested, he returns upright to vigorously peck at her nape. Even more intriguing is what happens *after* mating. When the brief copulation ends, the male often repositions himself to grasp the back of the female and then leans forward, wings flapping, until the female lets go of the perch. Both tumble in a downward spiral, locked together, for some distance until the male releases her and both fly away!

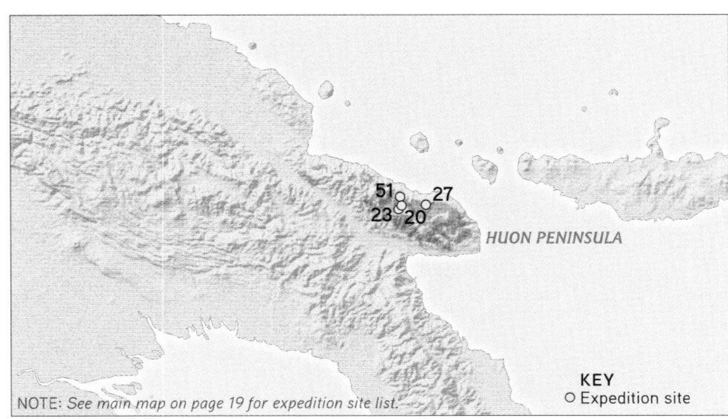

ELEVATION RANGE

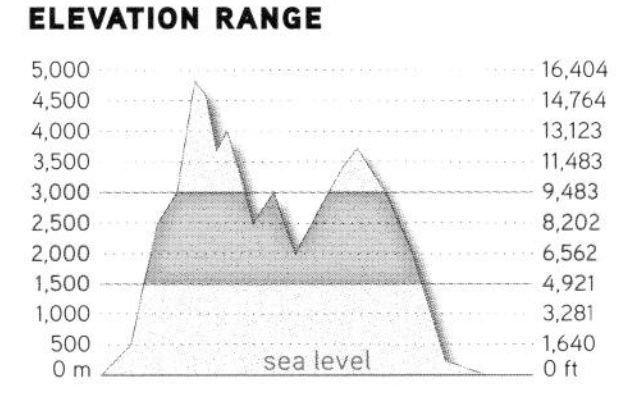

CONSERVATION STATUS LC

VOICE Little known; a throaty froglike croak repeated two or three times is given from display perches

Female feeds young. Site 17, 3 Oct 06

Male feeding. Site 17, 26 Oct 11

Adult male. Site 17, 26 Oct 11

STEPHANIE'S ASTRAPIA ~ *Astrapia stephaniae*

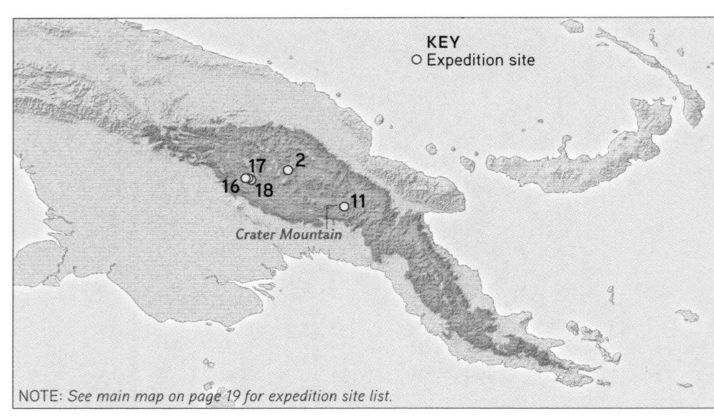

Stephanie's Astrapia, which was first described in 1884, gets its common name from Princess Stephanie, the wife of Crown Prince Rudolph of Austria-Hungary (who happens to be the namesake of the Blue Bird-of-Paradise, *Paradisaea rudolphi*). This species is found in the higher elevations of the easternmost part of New Guinea's central cordillera and overlaps extensively with a close relative, the Ribbon-tailed Astrapia, at the western end of its range. It closely resembles the Ribbon-tailed Astrapia, but lacks the forehead "pom-pom" and narrow white tail of that species. The most prominent feature of the male Stephanie's Astrapia is its extremely long tail, which averages about 65 centimeters (26 inches) long. The two central tail feathers are the longest, and they are highly sought-after ornaments in the traditional ceremonial headdresses of many surrounding highland cultures. In a few places, a single headdress boasts the feathers of over a dozen individual males (see "People and Plumes" on page 177).

Courtship displays of Stephanie's Astrapia are a little better known than many of the other species in the genus *Astrapia*. Nevertheless, specific details of the postures and movements used are still only generally described. This is one of the few species for which we didn't observe any courtship behavior. Instead, we encountered and documented it feeding on the fruit of schefflera trees. But from other sources, we know that courtship takes place in communal leks, with several males flying back and forth among five or six trees on the edge of a natural forest clearing. David Gillison provided a firsthand account from the remote regions of Crater Mountain, where the species is not routinely hunted for plumes. These leks may have upward of a dozen individual males, visited by several dozen female-plumaged birds at once. Finding one of these "super leks" is still high on our wish list and serves as an example of the many fascinating facts about birds-of-paradise still left to learn!

ELEVATION RANGE

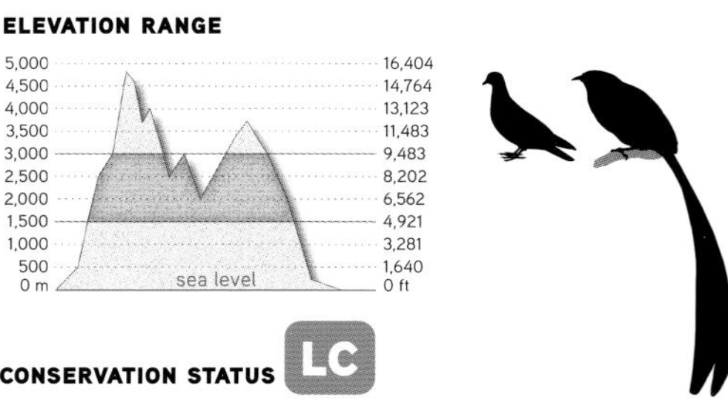

CONSERVATION STATUS LC

VOICE Variable; a loud, clear, whistled *waart-wheert* similar to that of the Ribbon-tailed Astrapia

---

Adult male chased by subadult. Site 1, 19 Aug 05

Adult male. Site 1, 12 Dec 10

Subadult male. Site 1, 10 Oct 05

RIBBON-TAILED ASTRAPIA ~ *Astrapia mayeri*

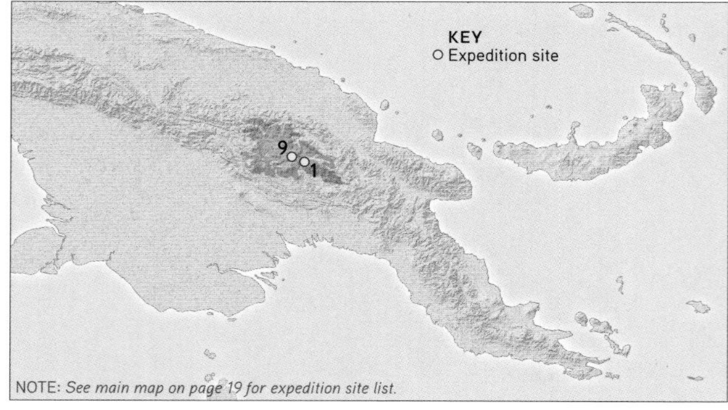

In 1950, S. Dillon Ripley, former secretary of the Smithsonian Institution, wrote, "The sight of a male flying off through the tree ferns in the jungle, long white ribbon feathers floating behind, must be like something from another world of fable and fantasy." He was right. Like the King of Saxony Bird-of-Paradise with its amazing banner-like wires on its head, or the Brown Sicklebill with its astonishing machine-gun voice, the Ribbon-tailed Astrapia is an otherworldly superstar. With tail feathers that can be more than three times its body length, the Ribbon-tailed Astrapia is a marvel of evolution. The species has the longest tail relative to body size in the entire bird world. Watching an adult male with a full-length tail fly or forage in the canopy is an experience that will never be forgotten. While probing for insects, its tail often becomes looped around branches. Incredibly, on more than one occasion, we've seen a male pause to turn back and use his beak to pull his looped tail toward him before flying away.

Found only in a tiny area of upper-mountain moss forests in western Papua New Guinea, the Ribbon-tailed Astrapia has always been one of the species most heavily hunted for decorative plumes in the traditional ceremonial dress of local tribes. While not uncommon within its range, it nevertheless is considered near threatened because of its very narrow distribution.

This species also has the distinction of being the latest bird-of-paradise species to be described. It was first seen alive by outsiders around 1936 by Jack Hides, a famous Australian patrol officer, who recognized its significance and collected several tail feathers. But they were somehow lost before they could be described. In 1938, Fred Shaw Mayer, the aviculturist and collector for whom the species is named, was given another set of tail feathers by a missionary while trying to locate the lost Hides feathers. Mayer sent the feathers to the British Museum of Natural History, and the species was finally described for science in 1939—about 28 years after the second-to-last bird-of-paradise described (the Short-tailed Paradigalla).

ELEVATION RANGE

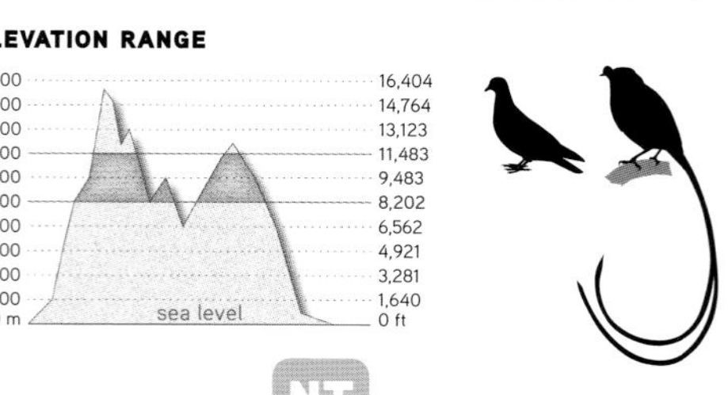

CONSERVATION STATUS NT

VOICE Loud, clear, whistled *woot-whit-whit* and variable throaty froglike croaks and growls

SEE ALSO PP 32, 34–35, 114–115, 161

Male displaying. Site 38, 31 Aug 09

Adult male. Site 38, 31 Aug 09

Adult male. Site 38, 27 Aug 09

SEE ALSO PP 4-5, 29, 98, 144, 173, 174

KING BIRD-OF-PARADISE ~ *Cicinnurus regius*

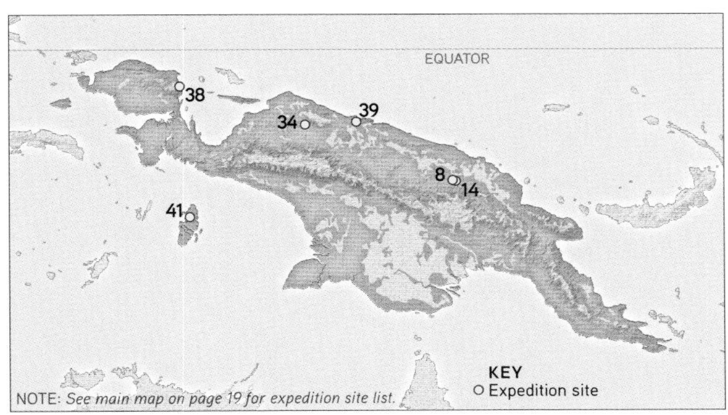

One of the two smallest in the family, the King Bird-of-Paradise is also among the most widespread bird-of-paradise species. Its diminutive size and preference for spending much of its time high in the forest canopy make it hard to spot in the wild. Nevertheless, it is a common and vocally conspicuous resident of lowland forests throughout all of New Guinea as well as several nearby islands. The King Bird-of-Paradise was first described by Linnaeus in 1758, and some regard it as the most "jewel-like" of all the birds-of-paradise because of the intense crimson-red color of the males. However, the species was well known to the Western world well before Linnaeus's time, as it was the only non-*Paradisaea* species illustrated in 1550 by Conrad Gesner, whose depictions were copied numerous times during the 16th and 17th centuries.

For the most part, males maintain their courtship display territories around a tangle of vines just under the canopy of large forest trees. They use both horizontal perches and vertical hanging vines for courtship displays. Perhaps the most extraordinary ornaments of the King Bird-of-Paradise are the two gemlike, iridescent emerald-green feather "disks" at the end of the male's long wiry tail. In the pre-mating display, in which the female perches adjacent to the displaying male on a horizontal branch, the male compacts his upper body, expands his special elongated iridescent green-tipped upper breast feathers around the head, and sharply cocks his tail upward so that while facing the female and bouncing up and down on outstretched legs, the two green disks at the end of the tail are bobbing and waving just above his head. Although this display usually takes place high in the canopy, often concealed behind twisted vines and leaves, it is nevertheless a spectacularly colorful and amusing sight.

ELEVATION RANGE

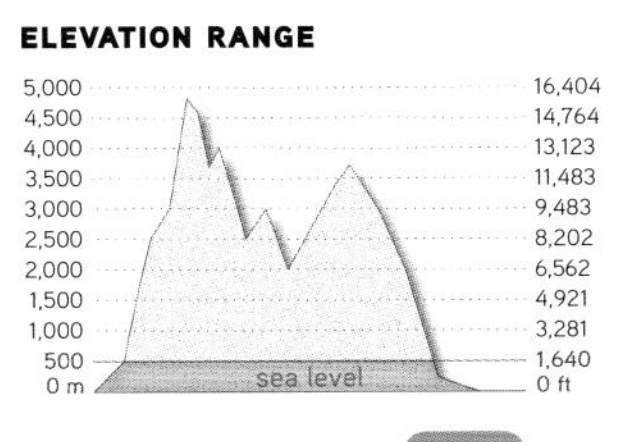

CONSERVATION STATUS LC

VOICE Variable nasal *kyer-kyer-kyer-kyer* or *qua-qua-qua*, sometimes throaty and other times more whistled

Adult male. Site 44, 2 Oct 10

Adult male. Site 6, 15 Nov 04

Display to female. Site 6, 20 Nov 04

SEE ALSO PP 30 82, 83, 168, 173

WILSON'S BIRD-OF-PARADISE ~ *Cicinnurus respublica*

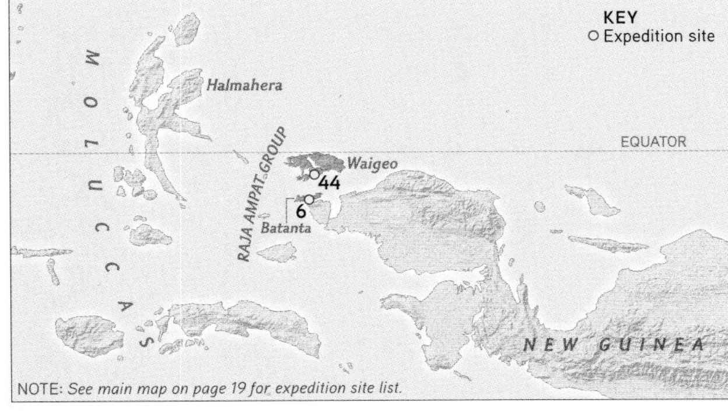

The Wilson's Bird-of-Paradise has the distinction of being the smallest bird-of-paradise, although measurements of the male overlap with its close relative the King Bird-of-Paradise. One of the most striking species, it appears as a riot of color with its intensely blue bare-headed skullcap, bright yellow nape, crimson back, and iridescent green breast shield. It occurs only on Waigeo and Batanta islands in the Raja Ampat Group. While regarded as primarily a hill-forest species, we observed several courts near sea level on Waigeo. Historically, this species may have had a broader elevational distribution that is diminishing due to growing human intrusions (agriculture and logging are substantial on both islands). Its IUCN Red List status is currently "near threatened," but given the species' very restricted distribution, a case for increasing that level to "vulnerable" could easily be made.

The species also has a colorful history surrounding its name. Prince Charles Lucien Bonaparte, nephew of Napoleon, first published its scientific name in February 1850 after seeing a male trade skin in Paris. Around the same time, English ornithologist Edward Wilson purchased the specimen and sent it to the Academy of Natural Sciences in Philadelphia. Six months later, American ornithologist John Cassin, unaware of Bonaparte's short note, named the new bird-of-paradise in honor of Wilson for his service to American ornithology. The rule of naming priority gave rightful status to Bonaparte's name, which referred to "that Republic which might have been Paradise had not the ambitions of Republicans . . . made their evil actions more like Hell." Yet Wilson is still recognized in the species' English name.

Along with its sister species the Magnificent, it is one of only two birds-of-paradise besides those in the genus *Parotia* that builds and maintains a terrestrial display court. Unlike the parotias, however, Wilson's doesn't display *on* the ground. Instead, like the Magnificent Bird-of-Paradise, it displays from small saplings near the ground while females watch from above on the same sapling.

ELEVATION RANGE

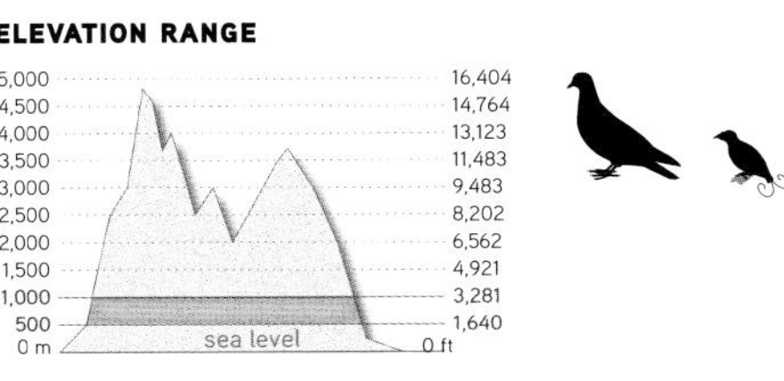

VOICE Loud, clear, powerfully whistled series *dyu-dyu-dyu-dyu-dyu-dyu-dyu*

Adult male. Site 22, 18 Dec 06

Adult male. Site 7, 28 Nov 04

Cleaning court. Site 7, 8 Dec 04

Display to female. Site 7, 8 Dec 04

SEE ALSO PP 10, 48, 135, 164

MAGNIFICENT BIRD-OF-PARADISE ~ *Cicinnurus magnificus*

A resident of hill and lower mountain forest throughout most of New Guinea, the Magnificent Bird-of-Paradise is the widespread main-island sister to the more limited Wilson's Bird-of-Paradise. It was first named by Thomas Pennant in 1781, but naturalist Edmé-Louis Daubenton had previously created an illustration in 1774. The Magnificent Bird-of-Paradise is closely related to the King Bird-of-Paradise, and they overlap extensively throughout much of their range. In some areas, the Magnificent and King occasionally hybridize despite the substantial differences in plumage and courtship behaviors. Because the Magnificent Bird-of-Paradise displays near the ground from a terrestrial display court and the King Bird-of-Paradise displays high in the canopy, it's hard to imagine how hybridization occurs. Still, the roughly two dozen hybrid specimens in museum collections are proof that they do. Interestingly, however, no hybrids have been spotted in the wild by naturalists.

Unlike the parotias, which prefer to build their courts on relatively level surfaces, male Magnificent Birds-of-Paradise typically put their courts on sloping—sometimes steeply sloping—ground. Their courts also often encompass rocky and uneven terrain. The reason for this difference is that, in contrast to parotias, Magnificent Birds-of-Paradise don't actually perform their courtship display *on* the ground, but instead perform them from vertical saplings *near* the ground. Males hop on the ground while clearing and maintaining their courts, but remain aloft when females are in attendance.

Highly impressive is their "horizontal cape display," in which the male leans back from a vertical sapling to become perpendicular to the perch. In this posture, the male's body is outstretched and its brilliant green iridescent breast shield is dramatically flattened along his front side, while his bright yellow nape-cape is flipped up to form a golden halo encircling the back of his head. The female watches all of this by looking down at him from just above on the sapling.

NOTE: *See main map on page 19 for expedition site list.*

KEY
○ Expedition site

ELEVATION RANGE

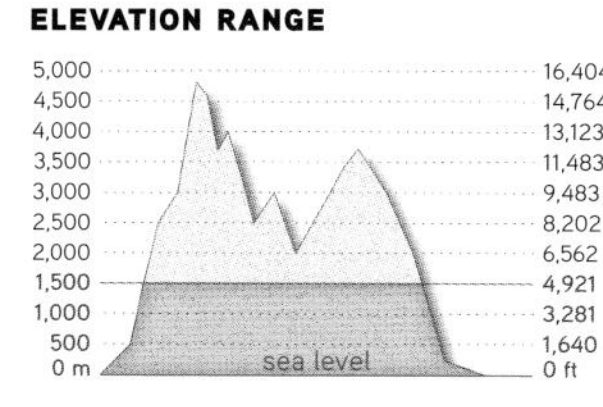

CONSERVATION STATUS LC

VOICE Repeated throaty *cheer-cheer-cheer-cheer-cheer* or *jeert-jeert-jeert-jeert*

Male display. Site 15, 18 Sep 06

Female plumage. Site 15, 14 Sep 06

Male calling. Site 15, 15 Sep 06

SEE ALSO PP 22, 54, 69, 140, 173

BLUE BIRD-OF-PARADISE ~ *Paradisaea rudolphi*

Because of its relative rarity and phenomenal coloration, the Blue Bird-of-Paradise is one of the most legendary of all. It has a limited distribution and occurs only in a narrow elevational range within the low- to midmountain forests of the eastern part of New Guinea's central cordillera. Because a substantial portion of the species' preferred habitat within the central valleys has been cleared for agriculture or otherwise disturbed by an increasing human population, the species is considered "vulnerable" on the IUCN Red List.

The Blue Bird-of-Paradise is by far the most distinctive member of its genus. In fact, it has frequently been placed in its own genus or subgenus of *Paradisornis*. The species was first discovered by Carl Hunstein in 1884 and described by Otto Finsch in 1885. Finsch named the species in honor of Crown Prince Rudolph of Austria-Hungary—the husband of Princess Stephanie, for whom Stephanie's Astrapia is named.

The most striking features of the Blue Bird-of-Paradise are its incredible blue plumage, bizarre upside-down courtship display, and pulsating courtship display song. The blue plumage of the male is nothing short of spectacular and appears translucent, almost glowing from within. Interestingly, the female Blue Bird-of-Paradise is much less dimorphic than in most other species. She also has blue coloration on the wings and tail, although it's not as brilliant as the male's. The courtship display of the male is also quite astonishing. The male displays from a horizontal or sloping branch in the forest understory. The display consists of the male fluffing out his flank feathers in front of his body to form a brilliant blue "apron" with a black-velvet oval shape in the center. He drops backward, tail first, to hang upside down for an extended period with his neck sharply bent so that his beak points upward. The female perches directly above and looks down at him while he hangs. During the display, the male gives an otherworldly pulsating electric-buzzing sound, which can go on for a considerable time without pausing.

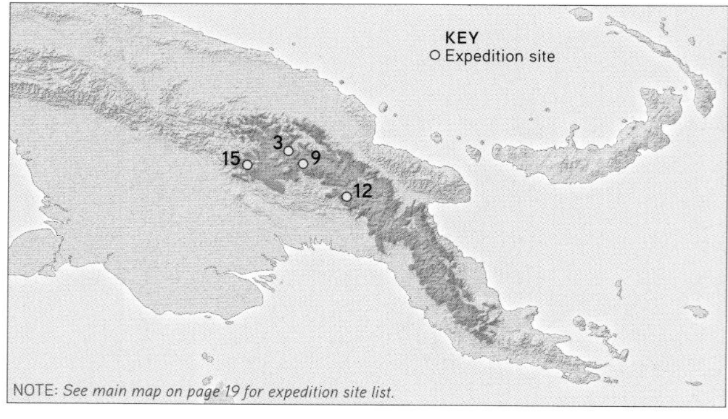

ELEVATION RANGE

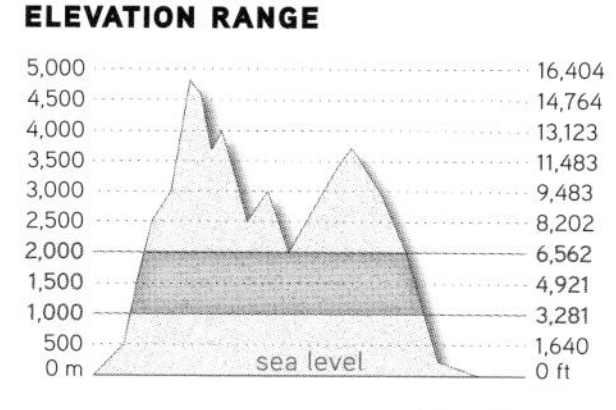

CONSERVATION STATUS VU

VOICE Nasal series of repeated *wah-wah* notes and also a more resonant *yaang-yaang-yaang*

Males at canopy display lek. Site 21, 8 Aug 07

Inverted display. Site 21, 7 Aug 07

Female plumage. Site 21, 7 Aug 07

SEE ALSO PP 44-45, 110

EMPEROR BIRD-OF-PARADISE ~ *Paradisaea guilielmi*

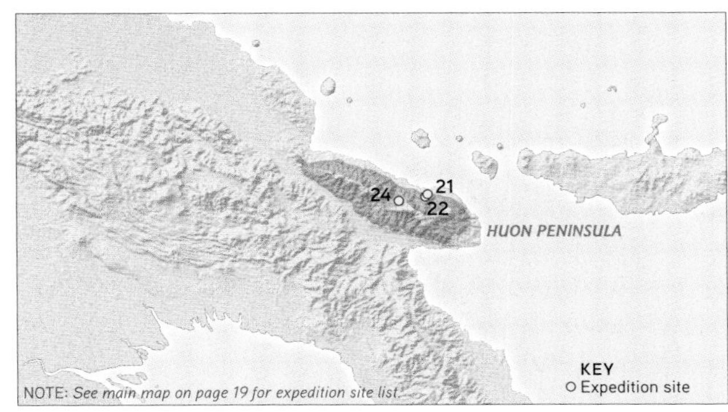

NOTE: *See main map on page 19 for expedition site list.*

Named in honor of Kaiser Friedrich Wilhelm III, Emperor of Germany and King of Prussia, the Emperor Bird-of-Paradise is the most distinctive member of the "typical" *Paradisaea* species group (i.e., the six *Paradisaea* species other than the Blue Bird-of-Paradise). Its very limited range is confined to the mountains of the Huon Peninsula in northeastern Papua New Guinea. The Emperor Bird-of-Paradise's preferred habitat is more exclusively midmountainous than all the other typical *Paradisaea* species; its range extends from the middle elevations down into hill forest, but not into the lowlands, as do the others.

The Emperor Bird-of-Paradise possesses a number of distinctive features. Compared to the other species, its unusual flank plumes are more sparsely filamented and whiter, and its characteristic pattern of iridescent green feathers covers most of the head and upper breast rather than just the face and throat. Most notably, its bizarre upside-down hanging courtship display is performed communally by multiple males in close proximity to one another within the canopy of the lek tree.

This species' courtship displays are poorly known and, to our knowledge, had never been photographed in the wild before this project. Most of what was known came from descriptions of birds in captivity. In 2007, when we documented this species for the project, Tim took the first photos of the Emperor Bird-of-Paradise displaying in the wild. Even watching this bird-of-paradise display, let alone photographing it, proved to be a huge challenge, because the display lek was located in the center of a very tall canopy tree with no nearby neighboring trees from which to view it. Tim deployed a remotely triggered camera hidden within the lek tree (see "The Leaf-Cam" on page 100) to obtain the incredible images of four adult males, in two pairs, hanging upside down at the same time. Although we didn't see a complete female visit, a female observes the display by standing above the male, more or less between his feet grasping the perch, to peer down and scrutinize his performance.

ELEVATION RANGE

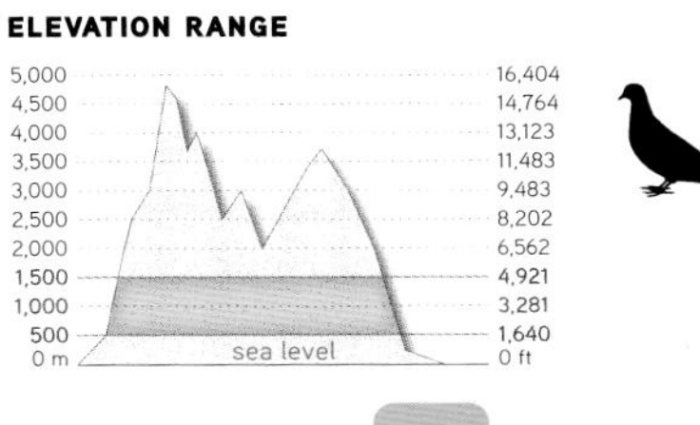

CONSERVATION STATUS

VOICE Repeated high-pitched and slightly nasal *whai-whai-whai* or *whaa-whaa-whaa*

Adult male at canopy display lek. Site 6, 23 Nov 04

Male calling. Site 43, 5 Oct 10

Female plumage. Site 43, 5 Oct 10

SEE ALSO PP 12, 91, 122, 173

RED BIRD-OF-PARADISE ~ *Paradisaea rubra*

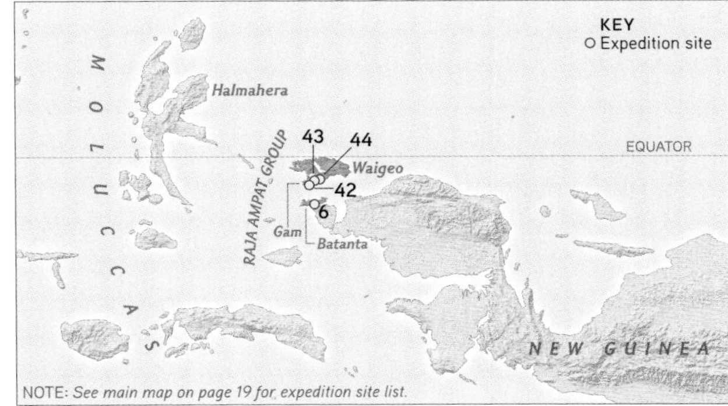

NOTE: *See main map on page 19 for expedition site list.*

Found only on the islands of Batanta, Gam, and Waigeo in the Raja Ampat Group west of New Guinea, the Red Bird-of-Paradise is an extraordinary bird that evolved in isolation far from its nearest relatives on the mainland. Several features make the appearance of this species quite distinctive, including the slightly elongated, iridescent emerald-green feathers above each eye, which form an unusual erectile facial ornament. Also, the bright red flank plumes are stiffer and hold their compact curved shape even when the male turns his head downward to become inverted during a display (in all the other *Paradisaea* species, the flank plumes are looser and cascade over the body when the male becomes inverted).

Perhaps the most extraordinary ornaments of the male Red Bird-of-Paradise are the two central tail feathers, which have been modified into ribbons with the look and feel of plastic that average 56 centimeters (22 inches) long. The surface of these appendages has become fused so that there are no discernible branches off the central shaft as in a typical bird feather. Other *Paradisaea* species do have elongated antenna-like tail wires, and the Blue Bird-of-Paradise has ribbonlike tail feathers, but the ribbons of the Red Bird-of-Paradise are unprecedented. In terms of structure, the only comparable feathers are the "head-wires" of the King of Saxony Bird-of-Paradise.

Like most of the *Paradisaea* species, this species adopts a head-down inverted display pose during its courtship display. The display is less static than the others', though, since the male constantly flutters his wings and moves from side to side. But it is only from this head-down posture that the unusual twisted shape of the tail ribbons makes sense. While inverted, the twisted segments help the ribbons to fall purposely to each side of the inverted male so that they perfectly surround his body in a heart-shaped frame (see photo on page 12-13).

ELEVATION RANGE

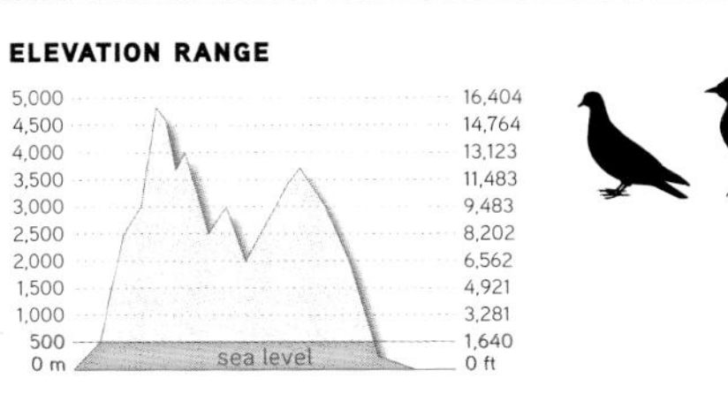

CONSERVATION STATUS NT

VOICE Loud, nasal *wok-wok-wok* or *werk-werk* that is repeated and variable in duration and tone

Males at lek. Site 4, 23 Sep 04

Female plumage. Site 4, 24 Sep 04

Adult male. Site 4, 22 Sep 04

GOLDIE'S BIRD-OF-PARADISE ~ *Paradisaea decora*

Another isolated species with a tiny distribution, Goldie's Bird-of-Paradise is found on only two islands, Fergusson and Normanby, in the D'Entrecasteaux Islands east of New Guinea. Scottish botanist Andrew Goldie discovered the species in the hills of Fergusson Island in 1882. Though believed to be primarily a hill-forest species living more than 300 meters (1,000 feet) above sea level, we documented a population in a lowland flood forest. While this could be an anomaly, it also begs the question of whether the species' range was once much more extensive. It could have been impacted by human-induced disturbances such as agriculture and, more recently, logging. Yet even when the species was first discovered, Goldie described encountering them primarily in the hills.

The female of Goldie's Bird-of-Paradise has a distinctive plumage pattern with slight barring on the undersides. While that marking is common in most female birds-of-paradise, it is not the norm within the genus *Paradisaea;* other than Goldie's, only females of the Blue Bird-of-Paradise have it.

The most extraordinary feature of the males, beyond the brilliant crimson-colored flank plumes, is their highly variable and dynamic vocal repertoire. Although males of most of the concentrated lekking species of *Paradisaea* vocalize at the same time, male Goldie's Birds-of-Paradise seem always to display in a pair and vocalize in what sounds like a duet. This duet is given not only during display but anytime the males are in the vicinity of their display area. In the peak of courtship displays, the males first vocalize somewhat idiosyncratically, but begin to synchronize as females draw nearer and then burst into their head-down display posture while performing the duet vocalization. It is extraordinary both visually and acoustically.

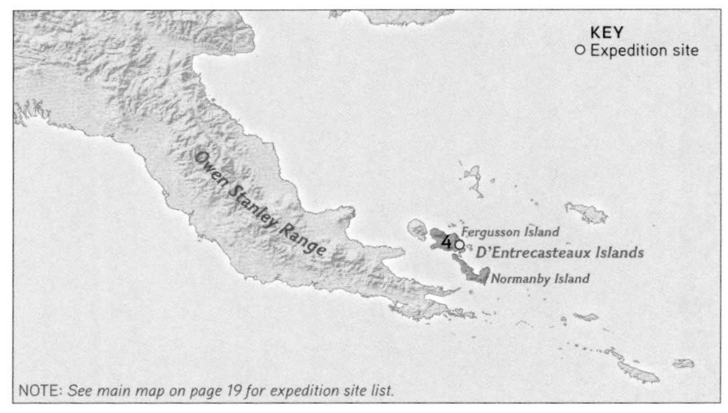

ELEVATION RANGE

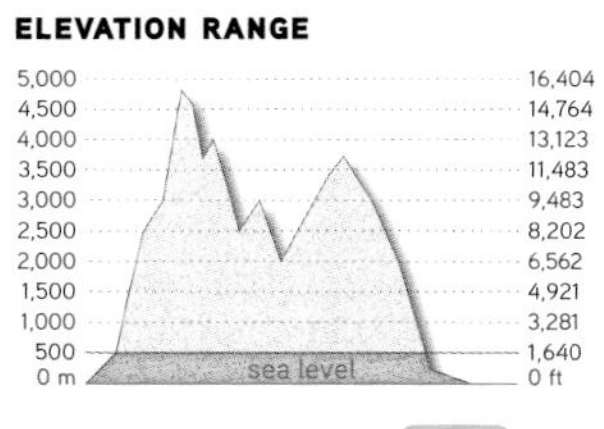

CONSERVATION STATUS NT

VOICE Loud, drawn-out *waah-waah-waah* and a variable loud *quong-quong-kwang-kwang* near the lek

SEE ALSO PP 86-87, 132-133, 136

Subadult male. Site 38, 25 Aug 09

Female plumage. Site 38, 20 Aug 09

Adult male. Site 38, 30 Aug 09

LESSER BIRD-OF-PARADISE ~ *Paradisaea minor*

This species is one of three (along with the Greater and Raggiana Birds-of-Paradise) that many people around the world, including many New Guineans, consider *the* bird-of-paradise. It is one of the most common within its range, which extends over a large part of northern New Guinea, from the Bird's Head Peninsula in the west to the Huon Peninsula in the east. Like its close relatives the Greater and Raggiana, the Lesser Bird-of-Paradise seems able to live in environments impacted by humans and can often be found around villages, settlements, and garden areas. Perhaps for this reason, it has also been the target of extensive hunting, both traditional and commercial, for the exquisite yellow-to-white flank plumes of adult males.

In July 1824, René Primevère Lesson, the naturalist onboard *La Coquille,* became the first Westerner to see a fully plumed bird-of-paradise in the wild, near the present-day Indonesian town of Manokwari. The species he saw while hunting for scientific specimens was the Lesser Bird-of-Paradise, and describing his encounter, he wrote:

> *The view of the first bird-of-paradise was overwhelming. The gun remained idle in my hand for I was too astonished to shoot . . . A* Paradisaea *suddenly flew in graceful curves over my head. It was like a meteor whose body, cutting through the air, leaves a long trail of light.*

Later, but before Wallace's account of the Greater Bird-of-Paradise, Lesson also noted that he thought the species might "form a harem after the manner of the Gallinaceous birds" and be "polygamist." Lesson was correct. Male Lesser Birds-of-Paradise perform courtship displays in traditionally used communal lek trees that contain upward of 20 birds in attendance. While we searched for one of these "mega-leks" of lore on several occasions, we never saw a display tree with more than a few adult males present at one time.

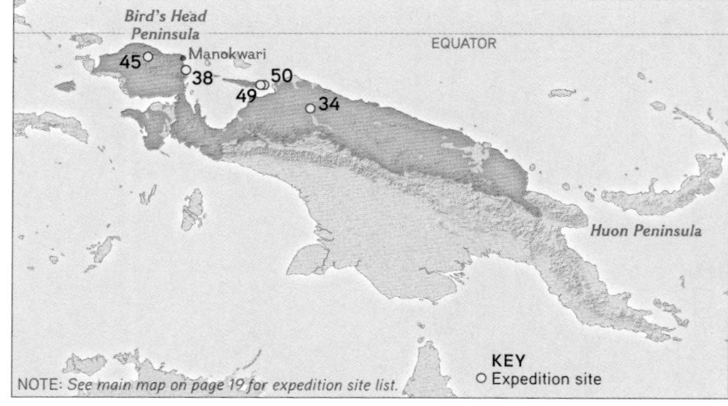

ELEVATION RANGE

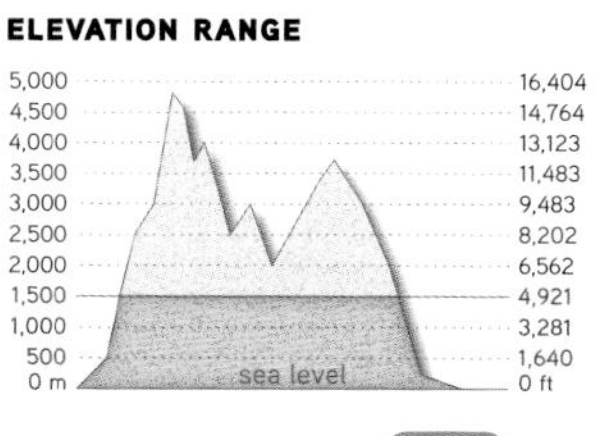

CONSERVATION STATUS LC

VOICE Loud, high-pitched, and somewhat nasal *wau-wau-wau* or *waah-waah* and a sharp *ki-ki-ki* from near the lek

SEE ALSO PP 40, 148

Female plumage. Site 47, 9 Dec 10

Male display. Site 47, 11 Dec 10

Adult male. Site 47, 10 Dec 10

RAGGIANA BIRD-OF-PARADISE ~ *Paradisaea raggiana*

The Raggiana Bird-of-Paradise is another of the iconic plumed birds-of-paradise widely recognized and cherished for both its beauty and its familiarity among the people who live within its range. It is found throughout nearly the entire eastern half of New Guinea, except the north coastal area west of the Huon Peninsula (where it is replaced by the Lesser Bird-of-Paradise). The range of the Raggiana Bird-of-Paradise thus encompasses most of the country of Papua New Guinea, which is why the species has been adopted as the national bird and emblem of that country. Its likeness is widely featured on currency, the flag, and countless products for both export and domestic markets.

First described in 1873, the species' unusual name comes from Marquis Francis Raggi of Genoa, who was a friend of Italian explorer Luigi D'Albertis. D'Albertis recognized a possible new species from trade skins he obtained and requested that the new species be named in honor of his friend. For many years, its status as a distinct species was questioned because of its overall similarity to and, where they commingle in the southwestern part of this species' range, known hybridization with the Greater Bird-of-Paradise. Now, however, all authorities consider it a separate species.

As one of the better known birds-of-paradise, the Raggiana's behavior and foraging habits have been fairly well documented, and its mating behavior has been the subject of several research projects. Males display from traditional communal lek trees, which include from two to ten individuals. Courtship behavior includes numerous poses and postures that accentuate the long red flank plumes of the male. The most extraordinary of these is the "static" or "flower" pose, in which the male turns away from the female with his head downward so that the flank plumes cascade like flowers over his back. Females observing this pose from behind are nearly engulfed by the plumes as they get close to the displaying male.

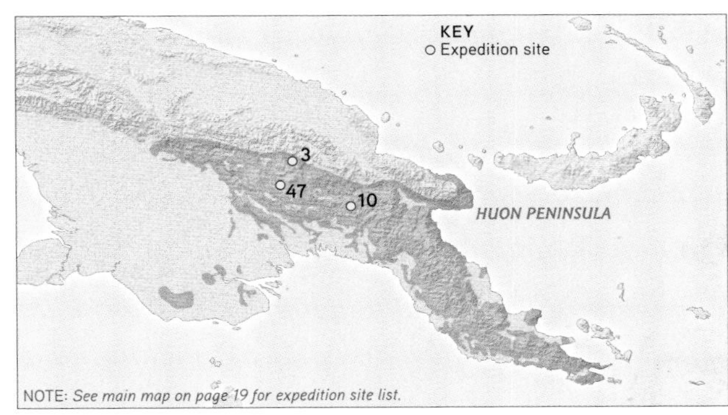

ELEVATION RANGE

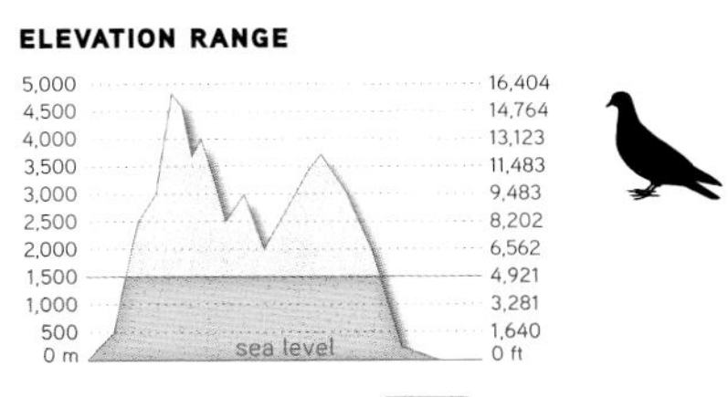

CONSERVATION STATUS LC

VOICE Loud *wau-wau-waauu-waauu* that increases in intensity and pitch, and also many variations of the *wau* notes

SEE ALSO P 145

Male shakes plumes at canopy lek. Site 41, 21 Sep 10

Adult male. Site 41, 16 Sep 10

Mating pair. Site 41, 19 Sep 10

GREATER BIRD-OF-PARADISE ~ *Paradisaea apoda*

Arguably the most legendary of all the plumed birds-of-paradise, the Greater Bird-of-Paradise remains surprisingly little known. This species was the first bird-of-paradise revealed to the Western world, through the ornamental trade skins presented to Magellan's skeletal crew as gifts to the King of Spain by the Sultan of Bachian (Bacan) during the famous circumnavigational voyage (see "Birds From Paradise" on pages 36-37). Linnaeus, in 1758, christened the Greater Bird-of-Paradise with the species name *apoda*, which means "footless" or "legless," in what appears to be a tongue-in-cheek reference to the widespread belief that birds-of-paradise, literally denizens of a heavenly paradise, lacked lower limbs. While Alfred Russel Wallace famously sought to collect this species in the Aru Islands, he became the first Westerner to see the spectacular group courtship displays of the males in a lek tree in the wild (see "In the Footsteps of Wallace" on pages 58-61).

Found in the Aru Islands and in the western half of southern New Guinea, the range of the Greater Bird-of-Paradise is quite a bit smaller than that of its two close relatives, the Lesser and Raggiana Birds-of-Paradise. This narrower distribution in a part of New Guinea that is less populated and less traveled, even today, helps explain why the Greater Bird-of-Paradise is less well known scientifically.

Much of what is known about their courtship displays derive from studies of a free-living, but greatly displaced and virtually captive, population introduced to Little Tobago Island in the West Indies by a well-intentioned conservationist, Sir William Ingram, in 1909. The species managed to survive for about 75 years, and during that time American ornithologist James Dinsmore studied them. The display behaviors he described overlap broadly with those known from the Lesser and Raggiana Birds-of-Paradise. These include the stunning head-down "static" pose in which the male turns away from the female on a downward-sloping branch so that his yellow plumes cascade over his back in an explosion of color.

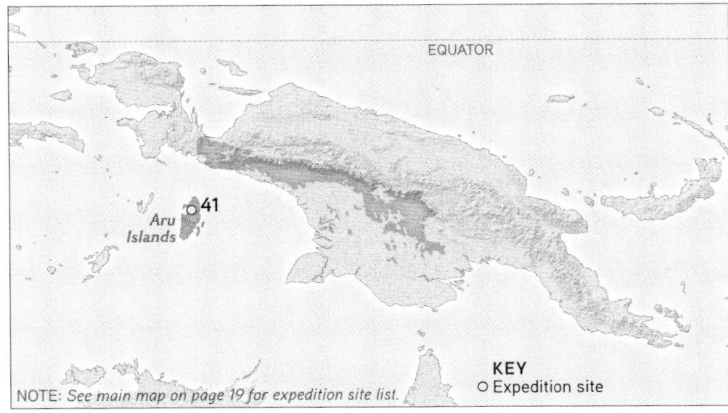

ELEVATION RANGE

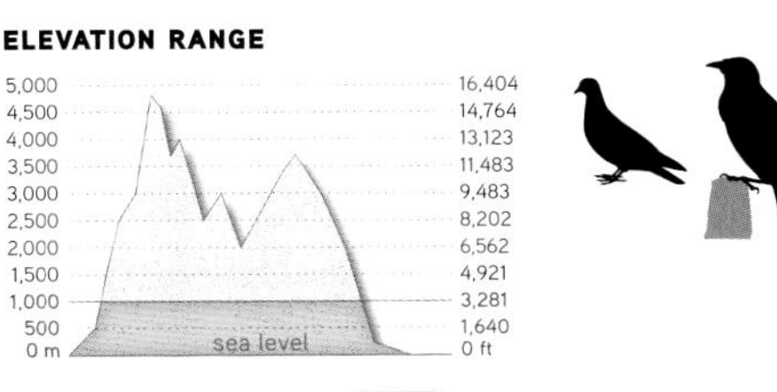

CONSERVATION STATUS LC

VOICE Loud *wauk-wauk-wauk* or *wau* with many variations, but usually given in a series of four or five notes

SEE ALSO PP 2-3, 24-25, 59, 100, 112, 131, 162, 172

Notes on Photography

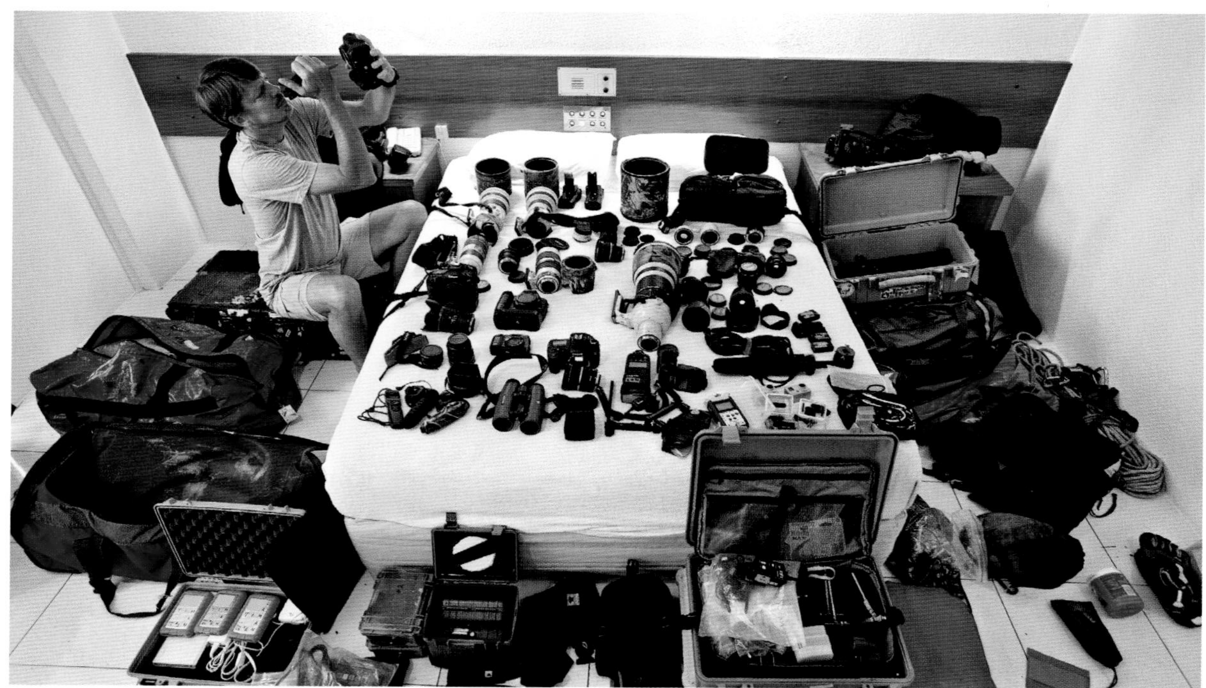

Lae, Papua New Guinea, 18 October 2011

Back in town after three weeks of camping in the rain forest, I spread out my camera gear on the bed in my air-conditioned hotel room to dry everything out. This is the bulk of my equipment for a trip shooting birds-of-paradise. Lenses ranging from wide-angle to 600 mm, four camera bodies, audio recording gear, tripods, lighting gear, hard drives, backpacks and waterproof cases for carrying everything into the forest, and duffels full of camping and tree-climbing gear. Not in the shot: two laptops; a duffel with boots, clothes, medical kit and personal gear; and lots of cables for remote shooting.

This is one frame from a time-lapse. To watch Tim pack up all his gear, see the time-lapse at *http://TimLaman.com.*

It is virtually impossible to walk around a rain forest carrying a camera and get a picture of a bird-of-paradise. They are just too high up in the trees, or too hard to approach closely. There is not a single bird image in this book that I made that way. To photograph birds-of-paradise, my approach was to find a place that they frequented and then wait for them to return. These sites were of two basic types: display sites and feeding sites. For most species, the ultimate goal was to photograph males displaying to females, so our focus was to locate display sites. This required us to spend lots of time in the field scouting. Once we had located a display site, the next step was to build a blind, either a ground blind or a much more difficult canopy blind. For the species that did not have regular display sites, I used the second option: We looked for a fruiting tree where we had seen birds feeding and staked it out. The final requirement for success was patience, often a lot of it.

Another approach I used was to put the camera into hidden positions closer to the birds, with unique angles or viewpoints I could never get from blinds. As detailed on pages 126-127, this remote camera approach required meticulous preparations and lots of special equipment, but it was ultimately successful and yielded some of my favorite photographs from the project.

It is an exciting time to pursue wildlife photography with all the tools we now have available. As a scientist, I value the ability to use these photographic tools to document the natural world, while my artistic side appreciates how imagery can be used to inspire a greater appreciation of nature. For me, the birds-of-paradise project has been an ideal blend of both.

—TL

Resources

Additional Reading

Selected books and *National Geographic* articles about birds-of-paradise

Coates, B. J. *The Birds of Papua New Guinea Including the Bismark Archipelago and Bougainville.* Vol. 2. Brisbane, Australia: Dove, 1990.

Forshaw, J. M., and W. T. Cooper. *The Birds of Paradise and Bower Birds.* Sydney: Collins, 1977.

Frith, C. B., and B. M. Beehler. *The Birds of Paradise: Paradisaeidae.* Oxford, England: Oxford University Press, 1998.

Frith, C. B., and D. W. Frith. *Birds of Paradise: Nature, Art, History.* Malanda, Australia: Frith and Frith, 2010.

——. "Family Paradisaeidae (Birds-of-paradise)," in *Handbook of Birds of the World.* Vol. 14. *Bush-shrikes to Old World Sparrows*, edited by J. Del Hoyo, A. Elliott, and D. A. Christie, 404-492. Barcelona: Lynx Edicions, 2009.

Gilliard, E. T. *Birds of Paradise and Bower Birds.* London: Weidenfeld and Nicolson, 1969.

——. "New Guinea's Rare Birds and Stone Age Men." *National Geographic* 103 (1953): 421-88.

Holland, J. S., and T. Laman. "Feathers of Seduction." *National Geographic* 212 (2007): 82-101.

Marshall, A. J., and B. M. Beehler. *The Ecology of Papua.* Parts 1 and 2. Singapore: Periplus, 2007.

Ripley, S. D. "Strange Courtship of Birds of Paradise." *National Geographic* 97 (1950): 247-78.

Wallace, A. R. *The Malay Archipelago: Land of the Orangutan and Bird of Paradise.* New York: Macmillan, 1869.

For additional scientific resources, visit *birdsofparadiseproject.org*.

Conservation and Research

To help conserve and advance the scientific understanding of the birds-of-paradise and their habitats, consider supporting one of these organizations.

Birdlife Indonesia: Works to protect birds and conserve habitat across Indonesia, including special efforts on the island of Halmahera, which is the exclusive home of the Standardwing Bird-of-Paradise and Paradise-crow.

Conservation International: Works in both Indonesia and Papua New Guinea at multiple levels to protect nature for the benefit of humanity.

Cornell Lab of Ornithology: The mission of the Cornell Lab is to interpret and conserve the Earth's biological diversity through research, education, and citizen science focused on birds. Member support helps make projects like the Birds-of-Paradise Project possible.

Green Capacity: Works in Papua New Guinea to resolve conservation challenges through field research, training, and mentoring of national conservation professionals.

National Geographic Society: Since 1888, National Geographic has inspired people to care about the planet through exploration and discovery, bringing wonders like the birds-of-paradise to light.

Papua New Guinea Institute for Biological Research: Founded by dedicated Papua New Guinean scientists, the PNGIBR conducts research and training programs that integrate traditional knowledge and customs with modern concepts of conservation.

Raja Ampat Research & Conservation Center: Promotes conservation-minded, community-based tourism in the spectacular islands of the Raja Ampat Group, which is home to two birds-of-paradise species found nowhere else.

Tree Kangaroo Conservation Project: Fosters wildlife and habitat conservation and supports local community livelihoods in Papua New Guinea, especially in the Huon Peninsula region, which is home to three endemic birds-of-paradise (see page 110).

Travel

If you plan to visit the land of the birds-of-paradise, we can recommend these companies, with which we have worked personally.

Kasoar Travel: A new endeavor by Loïc Degen and Miguel Garcia to create conservation-minded tourism involving local communities in the Aru Islands.

Kumul Lodge: An excellent location to see several high mountain species of birds-of-paradise near Mount Hagen in Papua New Guinea.

Papua Bird Club: Organizes birding tours involving expert local guides like Zeth Wonggor, and gives back a lot to rural communities. Founders Kris Tindige and Shita Prativi were our primary scouts and expedition organizers in Indonesian New Guinea.

Papua Divers: Besides great diving, offers side trips to see Red and Wilson's Birds-of-Paradise in the Raja Ampat islands.

Trans Niugini Tours: Operates Ambua, Karawari, and Rondon Ridge lodges in Papua New Guinea, all of which are excellent places to see birds-of-paradise.

Internet Resources

The Birds-of-Paradise Project *(birdsofparadiseproject.org)*: Our website for all things bird-of-paradise. Follow future projects, learn more about the birds-of-paradise, find out how to help support our efforts to learn more about the birds-of-paradise, and much more.

Macaulay Library *(macaulaylibrary.org)*: The scientific archive for all our bird-of-paradise video and audio recordings at the Cornell Lab of Ornithology. Most of our 2,000+ recordings are available for online viewing by searching or browsing the ML website.

Acknowledgments

We would like to thank all the people and organizations that have helped us carry out the Birds-of-Paradise Project and bring it to fruition. Although we are taking a serious risk of leaving out somebody, we would like to mention as many of you as we can:

Project Research and Development

Over the years we've relied on many people for help in the various stages of research and development needed to undertake a project of this size and scope. For their special expertise and encouragement, we'd like to thank Bruce Beehler, Clifford and Dawn Frith, Andy Mack and Deb Wright, Mary LeCroy, Lisa Debek, Bret Benz, Ian Burrows, Phil Gregory, David Gillison, Peter Alden, and Thane Pratt. TL gives special thanks to Scott Edwards for providing an academic home base in the Harvard Museum of Comparative Zoology. ES thanks his academic advisors and colleagues, especially Rick Prum (University of Kansas/Yale), Town Peterson (University of Kansas), and Joel Cracraft (American Museum of Natural History).

Field Support

When you make so many expeditions to so many places, you rely on a lot of assistance from many people. First of all, we want to thank all the villagers and landowners, too numerous to name here individually, who generously allowed us to work on their lands, and who often worked for us as guides and provided field support. The main villages we worked out of were: Syoubri and Kwau in the Arfak Mountains; Wailebet in Batanta; Kama in Wapenamanda; Sebutuia on Fergusson Island; Dobu Village on Dobu Island; Upper Heroana at Crater Mountain; Tigibi in the Tari area; Gatop, Satop, Kotet, Towet, Yawan and Apalap, all in the Huon Peninsula; Papasena and Kwerba in the Mamberamo region; Butu Putih, Foli, and Labi-Labi in Halmahera; Oransbari and Senopi, both in the Bird's Head Peninsula; Nimbokrang in the Jayapura area; Wakua in the Aru Islands; Sawinggrai and Yenbeser on Gam; Waiwo and Saporkren, Waigeo; and, finally, Keranui and Barawai on Yapen.

In Indonesia, the late Kris Tindige and his partner and wife Shita Prativi were our main organizers for six very productive expeditions, and working with their local partners—Zeth Wonggor in the Arfak Mountains, Jamil in Nimbokrang, and others—was essential to our success. Max Ammer of Papua Diving provided great logistic support and the spectacular opportunity to fly with him and shoot aerials from his ultralight floatplane in the Raja Ampat islands. For our trip to Halmahera, David Purmiasa of Birdlife Indonesia was our key point person, and local guide Demianus Bagali, better known as Anu, got us to the field sites. In the Aru Islands, Loïc Degen kindly put us in touch with the men of the Karey clan from Wakua Village, who allowed us to photograph in their protected forest. Budi Iraningrum of Conservation International Indonesia was indispensable in securing our permits to join a research expedition to the Foja Mountains, jointly sponsored with the Indonesian Institute of Sciences (LIPI), to whom we are also extremely grateful.

In Papua New Guinea, we also received support from many people and would like to acknowledge: the Bates family, Bob, Pam, and Michael, and all their staff at their lodges at Karawari and Ambua, including especially guides Chris, Joseph, Benson, Peter, and manager Dobaim. At Kumul Lodge, Kim and Paul Arut and their whole staff, and especially guide Max. In Alotau and Fergusson Island, Conservation International's David Mitchell and the landowners of Sebutuia. At Gatop, Pastor Gerhard Schuler, who kindly opened his home to us. In Lae, Lisa Dabek and the whole Tree Kangaroo Conservation Project staff, especially Karau Kuna, Ruby Yamuna, Toby Ross, and Zachary Wells, who helped us organize trips to the Huon Peninsula. And thanks also to Polly Weissner for helpful information.

In Australia, Clifford and Dawn Frith generously provided their expertise and a base of operations for which we will be forever grateful, and we would also like to thank Roger McNeill, and Rollie and Liane Callow for their help.

Cornell Lab of Ornithology

We offer sincere thanks to Director John Fitzpatrick, the Program Directors, and the Administrative Board for the generous support at many levels that allowed us to see the project to fruition, as well as the Lab's many dedicated staff who've helped in numerous ways. Special thanks go to John Bowman and the Lab's multimedia team, especially Eric Liner and Ian Fein, for all the time and effort spent in the office, studio, and field on our behalf; former and current Macaulay Library (ML) directors, Jack Bradbury and Mike Webster, Curator of Audio Greg Budney, Assistant Curator of Video Ben Clock, and all the ML staff for supporting the growth of the project and putting up with an often absent Curator of Video(!); Evaristo Hernandez-Fernandez, Joanne Avila, and Diane Tessaglia-Hymes for the species silhouettes; Ben Freeman and Alexa Class for scouting in the Huon; and also Dan Baldassarre and field crew for Australian scouting.

Project Funding

Fieldwork in remote New Guinea is expensive, and we would like to acknowledge vital grant support from these organizations: the National Geographic Society (both the Committee for Research and Exploration and the Expeditions Council), the Cornell Lab of Ornithology, and Conservation International, with special thanks extended to President Russell Mittermeier.

National Geographic Society

We would like to thank President Tim Kelly for taking a special interest in this project and helping it to become the Society-wide project it has become, and extend our gratitude to the many people from across the Society who have embraced and supported this project. John Francis and Rebecca Martin deserve to be especially singled out for their early and enthusiastic support that helped make this project possible.

At *National Geographic* magazine, TL would like to thank editors Bill Allen and Chris Johns, and directors of photography Kent Kobersteen, David Griffin, and Kurt Mutchler, who were all involved in various stages of this long project, for their vital support. TL especially thanks his photo editors John Echave and Kathy Moran. John, your support and belief in me since 1995 enabled my whole career at NGM, and thanks for guiding my first bird-of-paradise article to completion in 2007. Kathy, your vision and enthusiasm have guided the second half of this project and helped make it a success. Thank you both. NGM writers Jennifer Holland and Mel White have also been terrific colleagues and helped us tell the bird-of-paradise story to the millions of NGM readers. Also a special thanks to photo engineer Kenji Yamaguchi for technical assistance.

We want to extend our gratitude to the team from National Geographic Books that helped bring this book to fruition. Barbara Brownell Grogan had the vision to see the potential in this project and get us started, and you have all been very supportive as the book grew into a bigger project than we all imagined. Thanks to Anne Alexander, Jonathan Alderfer, Garrett Brown, Susan Hitchcock, and the whole team, including Fernando Baptista for the fantastic art, Carl Mehler for the beautiful maps, Bobby Barr and his team for their care in preparing the images for the book, and most especially our brilliant designer and partner in creating this book, Sanáa Akkach, whose vision shines throughout.

Family and Friends

TL: I especially dedicate this book to my mother, Evon Laman, who first inspired my interest in nature. She passed away during my last expedition for this project, and didn't get to see it come to fruition, but I think it would have made her happy. To my wife and partner in life, Cheryl Knott, your loving support alone has made this project possible. Thank you from the bottom of my heart. To my children, Russell (age 11) and Jessica (age 8), thanks for being such terrific kids, such an inspiration, and being so supportive of your often absent dad. To my dad, Gordon, who took me on my first wilderness trip when I was 12 and inspired my love of wild places, I thank you and Mom for everything and your lifelong support of my pursuits. And thanks to my extended family, including Sam and Joyce Knott, for all their support during my many long absences. Also a special thanks to my assistant Rachel Woolfson, who keeps the office running smoothly. I couldn't have done it without you all.

ES: I give the deepest debt of gratitude to wife, friend, and colleague, Kim Bostwick; without your support none of this would have been possible! To Nolan (age 5) and Natalie (age 1), it's hard to believe that you've not known life without the Birds-of-Paradise Project in it; I'm so lucky to have such a wonderful family that has supported my long periods away from home—this book is dedicated to you! To my mother, Barbara; my sister, Kelly; and my entire extended family—thank you all so much for everything you've done to support me in countless ways over these years. To Chuck and Sue Bostwick, thanks for all the help, great meals, and weekend getaways!; to Scott, Sandy, Jack, and Abbie, I couldn't even count the number of times you've been there for my family. I'm tremendously grateful to have such generous and supportive neighbors. Thanks to longtime friend Troy Murphy for the formative years spent "biologizing" in places near and far—those experiences led me all the way to New Guinea.

Index

Boldface indicates illustrations.

Birds of Paradise
Tim Laman and Edwin Scholes

Published by the National Geographic Society
John M. Fahey, Jr., *Chairman of the Board and Chief Executive Officer*
Timothy T. Kelly, *President*
Declan Moore, *Executive Vice President; President, Publishing and Digital Media*
Melina Gerosa Bellows, *Executive Vice President; Chief Creative Officer, Books, Kids, and Family*

Prepared by the Book Division
Hector Sierra, *Senior Vice President and General Manager*
Anne Alexander, *Senior Vice President and Editorial Director*
Jonathan Halling, *Design Director, Books and Children's Publishing*
Marianne R. Koszorus, *Design Director, Books*
Susan Tyler Hitchcock, *Senior Editor*
R. Gary Colbert, *Production Director*
Jennifer A. Thornton, *Director of Managing Editorial*
Susan S. Blair, *Director of Photography*
Meredith C. Wilcox, *Director, Administration and Rights Clearance*

Staff for This Book
Garrett Brown, *Editor*
John Paine, *Text Editor*
Sanáa Akkach, *Art Director*
Carl Mehler, *Director of Maps*
Gregory Ugiansky, *Map Researcher and Production*
Judith Klein, *Production Editor*
Galen Young, *Rights Clearance Specialist*
Katie Olsen, *Design Assistant*

Manufacturing and Quality Management
Phillip L. Schlosser, *Senior Vice President*
Chris Brown, *Vice President, NG Book Manufacturing*
George Bounelis, *Vice President, Production Services*
Nicole Elliott, *Manager*
Rachel Faulise, *Manager*
Robert L. Barr, *Manager*

Photo Credits

All photographs by Tim Laman unless otherwise noted:

36, The Pierpont Morgan Library/Art Resource, NY; 37 (UPLE), Princeton University Library, Historic Maps Collection; 37 (UPRT), Réunion des Musées Nationaux/Art Resource, NY; 38, Universität Mannheim; 39 (UP and LO), Petrus Plancius, 1594/Wikipedia (http://en.wikipedia.org/wiki/File:1594_Orbis_Plancius_2,12_MB.jpg); 58, Art by T. W. Wood, from *The Malay Archipelago* by Alfred Russel Wallace, 1898, Macmillan and Co.; 59 (UPRT), From *The Malay Archipelago* by Alfred Russel Wallace, 1898, Macmillan and Co.; 60 (LO), Art by Keulemans, from *The Malay Archipelago* by Alfred Russel Wallace, 1898, Macmillan and Co.; 76, Art by Baines, from *The Malay Archipelago* by Alfred Russel Wallace, 1898, Macmillan and Co.; 89 (LE), From *The Malay Archipelago* by Alfred Russel Wallace, 1898, Macmillan and Co.; 91 (RT), Art by T. W. Wood, from *The Malay Archipelago* by Alfred Russel Wallace, 1898, Macmillan and Co.; 128, Edwin Scholes III; 150, Edwin Scholes III; 151 (LE), Edwin Scholes III; 153 (LOLE), Eric Liner/The Cornell Lab of Ornithology/Cornell University; 153 (LORT), Image Archive ETH-Bibliothek, Zurich; 195 (LE A, B, C, D), Edwin Scholes III; 197 (LORT and UPRT), Edwin Scholes III; 207 (UP B), Edwin Scholes III.

The National Geographic Society is one of the world's largest nonprofit scientific and educational organizations. Founded in 1888 to "increase and diffuse geographic knowledge," the Society works to inspire people to care about the planet. National Geographic reflects the world through its magazines, television programs, films, music and radio, books, DVDs, maps, exhibitions, live events, school publishing programs, interactive media and merchandise. *National Geographic* magazine, the Society's official journal, published in English and 33 local-language editions, is read by more than 40 million people each month. The National Geographic Channel reaches 370 million households in 34 languages in 168 countries. National Geographic Digital Media receives more than 15 million visitors a month. National Geographic has funded more than 10,000 scientific research, conservation and exploration projects and supports an education program promoting geography literacy. For more information, visit www.nationalgeographic.com.

For more information, please call 1-800-NGS LINE (647-5463) or write to the following address:

National Geographic Society
1145 17th Street N.W.
Washington, D.C. 20036-4688 U.S.A.

For information about special discounts for bulk purchases, please contact National Geographic Books Special Sales: ngspecsales@ngs.org

For rights or permissions inquiries, please contact National Geographic Books Subsidiary Rights: ngbookrights@ngs.org

ISBN: 978-1-4262-0958-1

Printed in China

13/RRDS/2